SENSE OF PLACE, HEALTH AND QUALITY OF LIFE

Geographies of Health

Series Editors

Allison Williams, Associate Professor, School of Geography and Earth Sciences, McMaster University, Canada
Susan Elliott, Dean of the Faculty of Social Sciences, McMaster University, Canada

There is growing interest in the geographies of health and a continued interest in what has more traditionally been labeled medical geography. The traditional focus of 'medical geography' on areas such as disease ecology, health service provision and disease mapping (all of which continue to reflect a mainly quantitative approach to inquiry) has evolved to a focus on a broader, theoretically informed epistemology of health geographies in an expanded international reach. As a result, we now find this subdiscipline characterized by a strongly theoretically informed research agenda, embracing a range of methods (quantitative; qualitative and the integration of the two) of inquiry concerned with questions of: risk; representation and meaning; inequality and power; culture and difference, among others. Health mapping and modeling, has simultaneously been strengthened by the technical advances made in multilevel modeling, advanced spatial analytic methods and GIS, while further engaging in questions related to health inequalities, population health and environmental degradation.

This series publishes superior quality research monographs and edited collections representing contemporary applications in the field; this encompasses original research as well as advances in methods, techniques and theories. The *Geographies of Health* series will capture the interest of a broad body of scholars, within the social sciences, the health sciences and beyond.

Also in the series

Therapeutic Landscapes
Edited by Allison Williams
ISBN 978 0 7546 7099 5

Sense of Place, Health and Quality of Life

Edited by

JOHN EYLES

ALLISON WILLIAMS

McMaster University, Canada

ASHGATE

Published by
Ashgate Publishing Limited
Gower House
Croft Road
Aldershot
Hampshire GU11 3HR
England

Ashgate Publishing Company
Suite 420
101 Cherry Street
Burlington, VT 05401-4405
USA

www.ashgate.com

British Library Cataloguing in Publication Data
Sense of place, health and quality of life. - (Ashgate's geographies of health series)
1. Environmental health 2. Environmental psychology
3. Clinical health psychology 4. Quality of life
I. Eyles, John II. Williams, Allison, 1965-
613.1

Library of Congress Cataloging-in-Publication Data
Sense of place, health, and quality of life / [edited] by John Eyles and Allison Williams.
p. cm. -- (Ashgate's geographies of health series)
Includes bibliographical references and index.
ISBN-13: 978-0-7546-7332-3 (hardback)
ISBN-10: 0-7546-7332-4 (hardback)
1. Environmental health. 2. Environmental psychology. 3. Clinical health psychology. 4. Quality of life. I. Eyles, John. II. Williams, Allison, 1965- III. Series
[DNLM: 1. Environmental Health. 2. Quality of Life. 3. Social Environment. WA 30 S478 2008]

RA566.S46 2008
613'.1--dc22

2008002419

ISBN 978 0 7546 7332 3

Mixed Sources
Product group from well-managed forests and other controlled sources
www.fsc.org Cert no. SA-COC-1565
© 1996 Forest Stewardship Council
FSC

Printed and bound in Great Britain by
MPG Books Ltd, Bodmin, Cornwall.

Contents

List of Figures

List of Tables

List of Contributors

Gregory Ashworth

Professor of Heritage Management, Department of Planning, Faculty of Spatial Sciences, R.U. Groningen, Rijksuniversiteit Groningen

Educated at Universities of Cambridge (BA.Hons. 1962: Economics and Geography), Reading (M.Phil. 1966: Economic development) and London (PhD.1974: Tourism planning and policy). Taught at Universities of Wales, Portsmouth and, since 1979, Groningen. Since 1994, Professor of Heritage Management and Urban Tourism in the Department of Planning, Faculty of Spatial Sciences, University of Groningen (NL). Main research interests include heritage management [*Tourist-Historic City* (Wiley, 1990); *Heritage Planning* (Geopers, 1992); *Dissonant Heritage* (Wiley, 1996); *European Heritage Planning and Management* (Intellect, 2001); *A Geography of Heritage* (Arnold, 2001); *Construction of Built Heritage* (Ashgate 2001); *Senses of Place: Senses of Time* (Ashgate) 2005)], tourism planning [*Marketing in the Tourism Industry* (Croom Helm, 1984); *Marketing Tourism Places* (Routledge, 1990); *Tourism and Spatial Transformation* (CABI, 1996); *Horror and Human Tragedy Revisited* (Intellect, 2005), and place marketing [*Selling the City* (Wiley, 1990); *Binnensteden* (Samson, 1998)] c.15 books for commercial publishers, c. 300 chapters, and numerous journal articles. Currently consulting on planning issues with cities of Belfast, Perugia, Montevideo and Malta Tourism Authority. Presently serving on the Boards of The Irish Academy of Heritage, and The Institute of Italian Landscape Management. Currently involved in various international academic research networks on strategic planning, place marketing, and heritage planning.

Michael Buzzelli

Assistant Professor, Department of Geography, University of Western Ontario

Michael Buzzelli's research interests centre on the conceptual and empirical modeling of the social and environmental determinants of health. He was a Principal Investigator on Health Canada's Border Air Quality Strategy, a set of studies in the transboundary region of BC/Washington state developing new methods of air pollution exposure analysis for epidemiologic studies. Dr Buzzelli received his PhD from McMaster and his current research is funded primarily by Health Canada and SSHRC. He was Chair of the Health and Society Programme at the University of

British Columbia (2004–06) and Queen's National Scholar at Queen's University (2006–07).

Carles Carreras

Professor of Human Geography Universitat de Barcelona

Teaching at the Human Geography Department of the Universitat de Barcelona since 1972, from 1986 as a Professor. Invited Professor at the universities of Napoli (Italy), Sao Paulo (Brazil) and Bryn Mawr College (United States). Author of a handbook on Human Geography (Ed. Universitat de Barcelona, Barcelona, 1998). Specialised in urban research, had published some general works (*La ciudad. Enseñanzas del fenómeno urbano*, Ed. Anaya, Salamanca, 1980; *Geografia urbana de Barcelona*, Ed. Oikos-Tau, Vilassar de Mar, 1993; *Universitat i ciutat*, 2000; *La Barcelona literària*, Ed. Proa, Barcelona, 2003), and some retail and consumption approaches (*Les àrees de concentració comercial de Barcelona*, COCINB, Barcelona, 1990; *Els eixos comercials metropolitans*, Ajuntament de Barcelona, Barcelona, 1999; *Atles comercial de Barcelona*, Ajuntament de Barcelona, Barcelona, 2003).

Lily DeMiglio

School of Geography and Earth Sciences McMaster University

Currently a graduate student in the School of Geography and Earth Sciences at McMaster University in Hamilton, Ontario, Canada. She earned an Honours Bachelor of Science degree in Psychology and a Bachelor of Arts degree in Health Studies at McMaster University. Her present graduate research in human and health geography, under the supervision of Dr Allison M. Williams, involves examining sense of place at a community level and how it varies based on age, length of residency and neighbourhood characteristics. Her broad research interests also include therapeutic landscapes, rural health disparities and environmental health issues.

John Eyles

University Professor, School of Geography and Earth Sciences, McMaster University

John Eyles received his PhD from the University of London, working at Reading University and Queen Mary and Westfield College in London, before McMaster University. He is a Distinguished University Professor, based in the School of Geography and Earth Sciences with cross-appointments in Sociology and Clinical Epidemiology and Biostatistics. He has authored or co-authored some 300 books, peer-reviewed articles and technical reports, working closely with provincial and national health and environment agencies. His current research interests include the social determinants of health, urban health, access to health care interventions and the role of scientific knowledge in public policy.

Christine Heidebrecht

School of Geography and Earth Sciences, McMaster University

Christine Heidebrecht hold a Bachelor of Arts in Anthropology and Health Studies from McMaster University and a Master's in Epidemiology and Community Medicine from the University of Ottawa. Her primary research interests are infectious disease, health equity, health systems and community health, with a focus on global health. She conducted her Master's research in South Africa, and has recently spent time in Zambia, engaged in health system strengthening work. She now works in Toronto, Ontario.

Marko Krevs

Assistant Professor, Department of Geography, University of Ljubljana

Marko Krevs received his PhD in Geography from the University of Ljubljana, and is currently an Assistant Professor in the Department of Geography, University of Ljubljana. His research interests include quantitative and qualitative methodology in geography, geoinformatics, social and cultural geography, geography of population and settlements, and geography of quality of life. Recent publications include *Geographic Information Systems in Slovenia* (with Perko et al., 2006) and *Perceptual Spatial Differentiation of Ljubljana* (*Dela 21*, 2004).

Lynne C. Manzo

Associate Professor, Department of Landscape Architecture, Affiliate Faculty, PhD Program in the Built Environment, Affiliate Faculty, Interdisciplinary PhD Program in Urban Design and Planning, College of Architecture and Urban Planning, University of Washington

Lynne Manzo is an Associate Professor in the College of Architecture and Urban Planning at the University of Washington in Seattle. She received her PhD in Environmental Psychology from the City University of New York. She specializes in the study of place meaning and identity, affordable housing, and the politics of place. Her work appears in such journals as Housing Policy Debate, Journal of Environmental Psychology, the Journal of Planning Literature, and the Journal of Architecture and Planning Research, and she has presented her work at the annual conferences of the Environmental Design Research Association, the Council of Educators in Landscape Architecture, and Urban Affairs Association, as well as at a number of universities and professional design firms.

Bruce Newbold

Professor, School of Geography and Earth Sciences, McMaster University

Bruce Newbold is also Director of the McMaster Institute for Environment and Health. He has a wide range of research interests including: immigrant settlement and acculturation issues in the US and Canada, aging and health, immigrant health, and environment and human health. He has published widely in these areas.

Helena Nogueira

Professor, Geography Department, University of Coimbra

Helena Nogueira earned her PhD at the University of Coimbra, where she has been teaching and researching since 1999. She is interested in health geography and her research is focused on determinants of health, health behaviours, neighbourhood effects on health, health inequalities and statistical methodologies applied to these issues. Helena has numerous publications in national and international reviews and she has been collaborating in several national and international projects. She has a premiated investigation about the maternal-child care in an imigrant population (Bial prize in Clinic Medicine 2006).

Michael E. Patterson

Associate Professor, College of Forestry and Conservation, University of Montana

Michael Patterson is an associate professor on the Wildlife Biology and Recreation Management in the College of Forestry and Conservation at the University of Montana. He received a Masters and PhD. from Virginia Polytechnic Institute and State University in 1993. His research interests focus on human dimensions of wildlife management, relationship to place, research logic and standards for knowledge, and science policy. He is especially interested in research exploring human interaction with nature and how these shape meanings and values; the way meanings and values shape how people interpret and respond to natural resource management controversies; and approaches to conflict resolution.

Edward Relph

Professor of Geography, Department of Social Sciences, University of Toronto Scarborough

Edward (Ted) Relph has written extensively on the topics of place, landscape and urban design. He is the author of *Place and Placelessness*, a phenomenological investigation that was among the first studies of place and has been referenced and excerpted by architects, psychologists, planners, sociologists, landscape architects, geographers, philosophers and others. His other books include *The Modern Urban Landscape*, and *The Toronto Guide* – a field guide that won a special award from the American Association of Geographers. His publications have been translated into Japanese, Chinese, Korean, and Portuguese. In 2004 he received the Distinguished Scholar Award from the Canadian Association of Geographers.

Paula Santana

Full Professor, Geography Department, University of Coimbra

Paula Santana is currently a Professor in-cathedra of Geography and a Researcher at the University of Coimbra. She is broadly interested in urban social geography but her research centres on determinants of health, health behaviour, environmental and place effects on health. In teaching, she combines aspects of health geography, urban planning, and environmental risk factors. She became associated with the IGU Commission on Health and Development in 1995 and was first elected to the steering committee in 2001.

Ingrid Leman Stefanovic

Professor of Philosophy and Director of the Centre for Environment, University of Toronto

Dr Stefanovic's teaching and research focus on how values and belief systems affect environmental decision-making. Projects have included design of an integrative management framework for a 30-member, interdisciplinary academic research team. In addition she has explored perceptions and attitudes toward the Lake Ontario Waterfront Trail, and has provided codes of ethics for park use. She has published many articles and several books, including *Safeguarding Our Common Future: Rethinking Sustainable Development* (SUNY 2000). A forthcoming book to be published by the University of Toronto Press is entitled *The Natural City: Re-Envisioning the Built Environment* and a book in progress addresses issues of environmental ethics.

David Streiner

Director of the Kunin-Lunenfeld Applied Research Unit, Toronto

David Streiner is also Assistant V.P., Research, at the Baycrest Centre; a Professor in the Departments of Psychiatry, Nursing, Social Work, Rehabilitation, and Public Health Sciences at the University of Toronto; and Professor Emeritus in the Departments of Clinical Epidemiology and Biostatistics and Psychiatry at McMaster University. With his colleague, Dr Geoff Norman, he has published four books, and he is the co-editor of three others. He has published over 260 articles, including a series, *Research Methods in Psychiatry*, in *The Canadian Journal of Psychiatry*, that now consists of 27 articles; and was one of the founding editors of *Evidence-Based Mental Health*. His real job, though, is a woodworker and cabinet maker.

Allison Williams

School of Geography and Earth Sciences, McMaster University, Hamilton, Ontario, Canada

Allison Williams completed her PhD at York University (Toronto, Canada) and, after holding permanent positions at both Brock University and University of Saskatchewan, is now an Associate Professor in McMaster University's School of Geography and Earth Sciences. Her background in social and health geography continue to contribute to interdisciplinary research teams, three of which are currently examining urban quality of life, informal palliative caregiving, and rural health service delivery. She has published over 50 refereed journal articles or book chapters.

Daniel R. Williams

Research Social Scientist, USDA Forest Service, Rocky Mountain Research Station, Fort Collins, Colorado

Research by Daniel R. Williams examining the impact of globalization on the meanings of natural landscapes has appeared in the *Journal of Environmental Psychology*, *Journal of Leisure Research, Society and Natural Resources*, *Norwegian Journal of Geography*, *Journal of Sustainable Tourism*, and *Environmental Management*. He recently co-edited *Multiple Dwelling and Tourism: Negotiating Place, Home and Identity*, published by CABI. He served as editor of the journal *Leisure Science* from 1993–1998 and in 1999 received a Fulbright Scholarship to study the meaning and use of Norwegian natural landscapes.

Preface

This collection brings together two sets of research interests for both editors. For Williams, it joins her interests in quality of life, therapeutic landscapes and health. For Eyles, it puts again front and centre earlier interests in sense of place more generally and health and quality of life. In the past couple of years we have begun to work together to try and establish the existence and types of relation between these complex and definitionally elusive ideas and realities. To help us with this, we brought together an international group of scholars for a week-long meeting in Hamilton, Ontario in summer 2007. All participants brought a first draft of a paper with them. These were presented and discussed in lively, collegial sessions as well as in the social program organized as part of the meeting. While most participants were geographers, there were planners, psychologists and philosophers present as well. After the meeting, all went home and reworked their papers based on workshop comment. On resubmission, all were reviewed not only by the editors but also by at least one other participant.

Gratitude is extended to all contributors, who worked diligently to submit their work in a timely manner and who incorporated editorial comments – large and small, with great generosity. We feel greatly rewarded to have worked with all our contributors and thank them for many insights shared through their work.

A very special thanks to Mirjana Vuksan, Lily DeMiglio and Christine Heidebrecht whose research assistance has immeasurably eased the job of editing this collection. They have worked diligently in the preparation of this collection, carrying the vast majority of responsibility for correspondence with all contributors, and giving the detailed attention required to formatting and style requirements. Their commitment to this project (including help with meeting preparation) is very much appreciated and will not be forgotten.

Heartfelt gratitude to all members of immediate family (Saeid, Sarah and Cameron) – all who have allowed the mental space and time to devote to this project, and who have ensured the balance needed for effective work (Williams). Thanks to immediate family for keeping perspective and to the vineyards of France and Italy for services to science (Eyles).

John Eyles
Allison Williams
Hamilton, Ontario

Acknowledgements

Special copyrights provided

Figure 10.2: Topophilia and topophobia of the neighbourhoods within the Municipality of Ljubljana (copyright permission granted by Dela 21).

Figures 10.3c, 10.3d: Mortality rate by 'NUTS4', Mortality in Slovenia (copyright permission granted by Miha Staut).

Dedicated to Dorothy from John

And to Saeid from Allison

Chapter 1

Introduction

John Eyles and Allison Williams

People simultaneously experience numerous risk and protective factors reflecting the sum total of natural, built and socio-cultural environmental factors (Lillie-Blanton and LaVeist 1996). We recognize that physical and social environments impact health, grounding risk in the complex mix of physical and social characteristics of places in and of themselves (Kearns 1995; Macintyre 1997), but know little about how different aspects of these environments interact in influencing health. One direction to understand these complex processes is to examine the mechanisms or pathways through which place and the social relations within it shape the health status of individuals and populations (Dunn and Hayes 2000; Hayes and Dunn 1998; Kaplan and Lynch 1997; Lynch and Kaplan 1997; Macintyre 1997; Blomley 1994; Syme 1994). Missing from current studies in the examination of health effects of local environments is the subjective meaning and importance that individual's give to where they reside, i.e their sense of place. We contend that sense of place is an important link in the pathway that translates population health determinants to health outcomes. The purpose of this edited collection is to attempt to shed light on this linkage. To frame this collection we begin by briefly outlining the population health discourse and specifically the role of environment and place within that. Here we will highlight the determinants that appear to be central to the link between sense of place and health (these will be the social, physical, economic and cultural context of health). Following this, the literature specific to sense of place is reviewed, particularly as it is associated with health. Finally an overview of the contributing chapters in the volume is provided.

The Determinants of Health

The health of populations has been a focus of policy attention for many decades. It dates perhaps most formally from the concern with living conditions in industrial cities in rapidly industrializing societies, especially the UK. Ashton and Seymour (1988) suggest that the public health movement with its emphasis on improving such environmental conditions as accessibility to potable water and clean air as well as the availability of sanitation, other waste management methods and unadulterated food lasted throughout most of the nineteenth century. It peaked in the 1870s to be gradually replaced by a more individualistic approach to improving health as germ theory and such developments as immunization and vaccination developed. Indeed the environmental changes were so successful that death rates fell and quality of life began to improve enormously. The success of such public or population-focused measures was highlighted by McKeown (1976) as was the great improvement in nutrition. Given this backdrop there could be greater emphasis on personal health practices and individualized clinical and medical interventions to improve health status.

It is not our intention to rehearse the history of public health or medical care. But from these descriptions we can see the factors that shape, influence and indeed determine health status emerging, e.g. environment, health care services, individual make-up. In the Canadian context, Mustard and Frank (1994) and Frank (1995) provide an outline and these factors, the outline being largely developed by the Canadian Institute of Advanced Research but based on developments in the policy arena dating from the 1974 Lalonde Report which identified four health fields, namely environment (physical and social), lifestyle, health services and biology. So CIAR argued that the health of populations is powerfully influenced by a nation's wealth-creating capacity and its distribution of wealth. Furthermore, individual

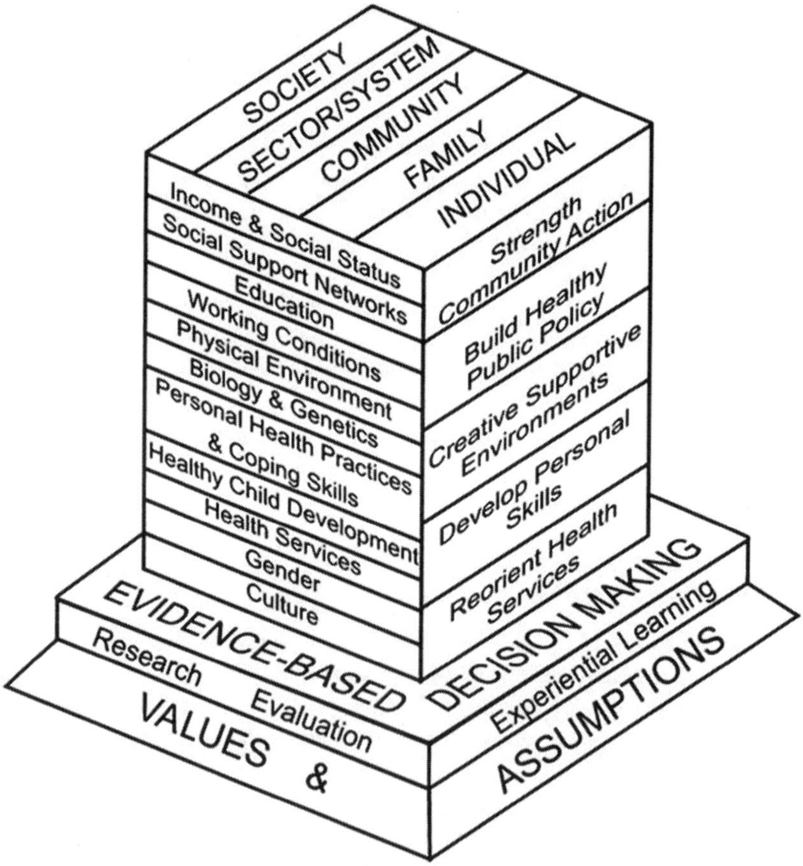

Figure 1.1 Population Health Promotion Model

Source: Adapted from Hamilton, N. and T. Bhatti (1996).

behaviours as well as such characteristics as age, gender, stage in the life cycle and immunoresponse system are also implicated.

Importantly the identification of these determinants of health was not only an academic pursuit. From the mid-1980s on, and specifically from the Ottawa Charter on Health Promotion, a policy and action lens has been present. Within the social determinants of health framework, emphasis has been placed on developing and promoting evidence-based decision-making , integrating policies that tackle several determinants and perhaps most importantly for our argument, promoting new approaches to policy, including the development of healthy public policy, partnerships and collaborations and community involvement and local action (PHAC 2002). These policy linkages have been seen in other jurisdictions as well (see Wilkinson and Marmot 2003 and Heymann et al. 2006). These linkages between science and action are vital for our project on senses of place and health. To help shape our argument we therefore utilize the population health promotion model (Hamilton and Bhattia 1996).

From this model, we may also observe that the determinants of health have been broadened since the earlier discussions of CIAR. PHAC now identifies twelve such determinants for which there exists an evidence base for their connection to the health status of populations. These are income and social status, social support networks, education, employment/working conditions, social environments, physical environments, personal health practices and coping skills, healthy child development, biology and genetic endowment, health services, gender and culture. While there is some differentiation of the determinants across developed nations, perhaps the most fundamental addition has resulted from WHO's participation in the discussions. WHO created a Commission on the Social Determinants of Health (CSDH) based on the continuing existence of health differences between countries and across the social gradient. Although reporting in 2008, CSDH has raised issues of equity, justice and empowerment (WHO 2007) and has focused on nine knowledge networks to derive evidence for policy and advocacy. These are globalization, health systems, urban settings, employment conditions, early child development, social exclusion, women and gender equity, priority public health conditions and measurement and evidence (see Vega et al. 2006). Thus, while there are strong parallels with PHAC's listing, there is a greater focus with WHO on context, equity and action. But we should note the strong linkages between their interests and those portrayed by the UN (2005) with their millennium development goals, namely to eradicate extreme poverty and hunger, achieve universal primary education, promote gender equity and empower women, reduce child mortality, improve maternal health, combat HIV/AIDS, malaria and other diseases, ensure environmental sustainability, and develop a global partnership for health. We shall explicitly link these goals and the determinants of health as well as the relations between health, equity and justice in some of the ensuing chapters.

Emphasis on action and equity requires too an emphasis on the evidence base for such activities. There is indeed strong research evidence for the existence of variations with respect to the determinants on health between populations at least at the aggregate level. There is of course a growing critical literature concerning the determinants of health (see Labonte and Schrecker 2007; Judge 1995; Muntaner

2004) and some concern about how population health status is measured by variables that best express the health of individuals. Thus, for example, personal health practices are often measured by such lifestyle variables as individual smoking status, physical inactivity, obesity and level of alcohol consumption (see Wilkinson and Marmot 2003). With respect to the topic of our collection – place and specifically sense of place – we may note that the environment has tended to be a neglected determinant (see Eyles 1999), partly because it is such a broad-based phenomenon, including housing, neighbourhood, toxic insults, and partly because it is very difficult to measure even in terms of individualized variables (e.g. treated as road traffic accidents or work-based injuries). It has thus, as expected, received critical treatment (see for e.g. Ben Shlomo et al. 1996; Mellor and Milyo 2001; Davey Smith et al. 2001). But we now turn to discuss explicitly the role of place as an influence in shaping the health of populations.

Geography, Place and Health

Considerations of geography or place with respect to any social phenomena, such as health, require recognition of the importance of scale. Gatrell (2002) points out that factors – determinants – important at one geographical level may have a non-existent or different effect at another scale. These issues have begun to interest health geographers and others (see Subramanian et al. 2001; Turrell et al. 2007), but are beyond the scope of our collection except as a constraint on place-health analyses. We are primarily concerned about individual perceptions of place; a place-based approach to health involves decisions about the definition and dimensions of place, the latter of which provide a context for individual's perceptions. Giddens (1985) and Agnew (1987) discuss components of place that have been incorporated into questions regarding health and place: locale (settings of interaction) and location (geographic area encompassing locale, defined by social and economic processes operating at a wider scale). Location also exists at other scales (e.g. the national level) and locales might include workplaces, living spaces (housing) or recreational places. Furthermore we note that there are various place characteristics which may affect place effects on health. Three are important: physical, e.g. the buildings, infrastructure and open spaces/nature; the social, e.g. the density of and interactions between inhabitants and their institutions; and psychological, e.g. identity and attachment to the physical and or social attributes of the locality. We suggest that the psychological dimension has often between underemphasized in studies, especially in so far as it relates to health. There has been much humanist interest in the meaning of place for well-being but little focus on variations in place meanings and what this entails for variations in health outcomes. As little work has been done to date on place meanings – i.e. senses of place – and health this edition tries to fill this gap by exploring the link in an interdisciplinary manner.

Sense of Place and Health

A significant body of theoretical and empirical studies discusses sense of place as an outcome of interconnected psychological, social and environmental processes in relation to physical place(s) (see DeMiglio and Williams this volume). Sense of place has been examined widely, spanning across a number of disciplines, such as environmental psychology, sociology and anthropology; it has been given a good deal of attention using a humanistic perspective in human geography, in terms of both the character intrinsic to a place as a localized, bounded and material geographical entity, and the sentiments of attachment and detachment that humans experience and express in relation to specific places (Relph 1976; Tuan 1974; Gesler 1992; Cresswell 1996; McDowell 1997; Oakes 1997). There seem to be some core elements to sense of place: rootedness, belonging, place identity, meaningfulness, place satisfaction and emotional attachment. While these comprise psycho-social constructs, recent work has recognised the importance of the physical environment as another core element of sense of place (Clark and Stein 2003; Pretty et al. 2003; Stedman 2003a; 2003b; Yung et al. 2002). Inherent in the sense of place construct is the core element of the physical environment. The focus on the relationships that people have with the physical environment is what makes sense of place distinct from concepts such as social capital or social cohesion.[1] Place-based research contributes to a better understanding of environment as a health determinant and may be applied to solutions directed at improving health, such as enhancing the health promoting qualities of the places where people spend their time (Macintyre et al. 1993). Furthermore health-based research that takes into account place as a determinant of health may enhance our understanding of both health and place. Our contributors try and provide through their case studies a nuanced understanding.

Systematic investigations of relationships between compositional (individually-based) and contextual (place-based) characteristics and health status have yet to determine fully the independent importance of contextual factors, as well as how they interrelate. The centrality of place to already known pathways between social circumstances and health (e.g. income inequalities and health, and social support and health) further reinforces the need for greater research on the role(s) of place in the social production of health.

Furthermore while studies of persons and place are increasing, most ignore a potentially important factor: individual's perceptions of and relationship to his/her place of residence, or their sense of place. Sense of place may be a key construct in place-based health research as it provides a conceptual link between the area-based

1 Drawing on the communicative theory of Habermas, Butz and Eyles (1997) interpret *sense of place* as being constituted socially and ideologically as well as ecologically. The underlying idea informing this interpretation establishes communication as being integrally associated with particular *sites* of interaction. The physical environment is the core element that "ties" *sense of place* irrevocably to the tangible: "Ecological aspects of the life are as much social and mental constructs as they are products of physical ecological settings, but they are nevertheless strongly rooted in that physical setting" (ibid. 11). As some researchers note, there is an innate link between who and where we are for "to be some*where* is to be some*one*" (Cheng et al. 2003, 90).

variables "out there, in place" and the internal biological processes and systems in individuals. Geographers have recently incorporated the construct of sense of place into health research. The work conducted on health and the meaning of place (Gesler 1992, 1993; Williams 1998, 1999; Kearns 1991, 1995) has directed attention to sense of place and its role in health; much of this work has used the therapeutic landscape concept, which recognizes sense of place as central to positive health experiences. We also have some evidence that sense of place contributes to community-level pathways and processes that positively influence health (Warin et al. 2000; Theodori 2001). The perceptions and meaning ascribed to place can also have negative connotations. For example, 'neighbourhoodism' is a term used to describe the stereotyping in media and public attitudes of poor, ghettoized neighbourhoods. The negative attributes of these areas are often attributed to people living there who, in turn, often internalize these features as being partly a reflection of their own lack of self-worth, leading to poorer health or quality of life (Williams et al. 2002). A positive image of neighbourhood, in turn, may have a salutary effect by enhancing personal attitudes, behaviours and self-concept, and thereby health and quality of life (Kearns et al. 2000; Meegan and Mitchell 2001; Healey 1998).

Health geographers have looked at related concepts such as the relationship between perceptions of specific environments and health. In James and Eyles' (1999) exploratory study of perceptions of health and the environment among men and women in lower and higher status areas, they found that both genders referred to health and environment as connected concepts. Men, however, referred to this connection more than women and lower income participants noted environment-related health problems more often but in less detail. In Wakefield et al.'s (2001) study of health risk perception and community action, one of their findings revealed that those interviewed tended to report adverse effects on their health if air pollution was visible.

A number of studies in disciplines other than health geography have examined the importance of how one perceives certain aspects of the local environment and have found significant effects on health. For example, Collins et al. (1998) found that among African-American mothers, perceptions of their residential environment (including police protection, personal safety, cleanliness and quietness) were associated with very low birth weight outcomes even after controlling for maternal behaviours such as alcohol use and cigarette smoking. Ewart and Suchday's (2002) application of the City Stress Inventory (CSI) found that CSI subscales were associated with elevated chronic levels of depression, anger, attitudes of interpersonal distrust, and low self-esteem. Similarly, in Fuller et al.'s (1993) examination of objective and subjective housing conditions and well-being, they concluded that objective housing conditions were not associated with many of their 10 measures of health but that housing satisfaction was significantly related to half of their measures of health.

Further advances in understanding the effects of place-based and individual-level factors on health are likely to be made by studies using primary data to investigate specific processes through which perceptions of place or area effects mediate health outcomes. These studies would need to incorporate direct measures of individuals' perceptions of place along with individual and place-based characteristics, and examine interactions such as the relation of the place-based socio-economic context to potential mediators of place effects. We recognize that sense of place is very likely a multi-dimensional (rather than uni-dimensional or bi-dimensional), and dynamic construct, in that the nature of the dimensions could vary over time and place as a function of the characteristics of an (a) individual, and (b) community, or potentially (c) both (a) and (b) working together. For example, the nature of a person's individual sense of place may be a complex combination of awareness of community characteristics and the individual's subjective reaction to those characteristics that he/she believes are salient community features. The subjective reaction could be positive, neutral or negative. Similarly, different community settings could necessitate differing dimensions, and the relative importance of those dimensions could also change based upon the salient features of the community, the surroundings of the community, and the amenities within the community. Recognizing the complexity of the sense of place and its relation to health, we put forward this collection as a beginning with respect to where sense of place is coming from and going to, how we might investigate its relation to health and a series of case studies which examine sense of place (variously conceived) on health (variously defined).

Overview of Book

The various chapters thus illustrate these complexities. DeMiglio and Williams provide a review of the sense of place and health literature, highlighting the different disciplines that have been involved in its study as well as the definitional complexities. Both Relph and Stefanovic in their chapters extend consideration of sense of place from its humanistic and philosophical underpinnings. Relph examines sense of place and well-being in relation to the environmental challenges of the present century pointing to cooperative ways of understanding and action. Stefanovic investigates the common elements in the philosophies of place and health, using an example of place-making among the young to illustrate the potential healthfulness of place. The next two chapters are primarily methodological. Eyles reviews and exemplifies different qualitative approaches to the study of sense of place and health. Given the uncertain, complex nature of these relations, it is likely that such approaches will provide a key conceptual, clarifying role. Williams et al. introduce one quantitative approach to sense of place-health studies using facet design to tease out the nature of sense of place constructs that can be used in health research.

The next series of contributions provide case studies of sense of place and health and well-being relations and demonstrate the complexity and variability of both. Manzo emphasizes home and housing in her study of the changing nature of low income housing in American cities. Using environmental psychological concepts such as place attachment and identity, she reports on how such change can reduce or militate against sense of place and well-being. Williams and Patterson use several cases pertaining to recreation in natural settings to argue for considering leisure and more generally time use as important dimensions of health and well-being. They also link the determinants of health to the millennial goals of improving quality of life in different ways in different settings. Time is also an element in Carreras' study of the senses of place of young people in Sarajevo. He reviews the changing nature of the images of this city which has experienced many political, economic and religious changes over time, with these layers being overlain but of different significance depending on age and generation. Thus while all images are in some ways relevant for everyday life and well-being, it is the desire of the young for consumerism and possible emigration that shapes their perspectives. Krevs reviews the situations in other fast-changing contexts, namely the post-socialist realities of Eastern Europe. He notes the changing quality of life in these settings before examining in more detail how quality of life and changing places intersect in Slovenia in particular. Santana and Nogueira use Lisbon as their case study to show how health and well-being, as measured by obesity, are shaped by various determinants of health. In fact, they show the relative unimportance of sense of place, measured in social terms, for this outcome. If their chapter highlights the contingent, perhaps irrelevant, nature of sense of place for health, the final two empirical chapters raise other points to consider. Buzzelli emphasizes the importance of inequities through considering environmental (in)justice in disadvantaged communities. While such communities may develop a resilience which is based on and indeed helps develop a sense of place, they may also have low levels of well-being and of individual health status. The relation between sense of place and health is not unidirectional or universal. In the final empirical chapter, Ashworth brings us back to place and the attempts to create place identity through heritage. He suggests that heritage is used to create an identity dividend and questions whether this can be maintained or even continue to exist if virtually all places are so regarded. Using Dutch policies as an example, he points to the problematic nature of the sense of place-well-being relation.

The contributions in this book tell therefore variable stories of the sense of place and health-well-being relationship. While it would seem intuitively correct to suggest that there is a relation between this psychological dimension of place and health, complexity gets in the way of easy formulations or answers. Much depends on how the terms are defined, the scale of analysis and the geographical and temporal context of the investigation. But in a concluding chapter, we try and go beyond the chapter contributions, reviewing not only these but the changing nature of place and health to point to ways to further enhance our understanding of this variable but intrinsically important relationship. As Liaschenko (1994, 19) notes, "places are symbolic constructions reminding us of our connections to others, to the natural world and animals, and to projects – they give meaning to our lives. Thought of in this way, we can see that place is important in shaping our identities and in fostering (or depleting) our sense of self and agency." Just as Peter (2002) does, after approvingly using this citation, our task is to continue to relate these senses to well-being and health.

References

Aday, L. (1989), *Designing and Conducting Health Surveys* (San Francisco: Jossey-Bass Publishers).

Agnew, J.A. (1987), *Place and Politics: The Geographical Mediation of State and Society* (Boston: Allen and Unwin).

Antonovsky, A. (1987), *Unravelling the Mystery of Health: How People Manage Stress and Stay Well* (San Francisco: Jossey-Bass Publishers).

Ashton, J. and Seymour, H. (1988), *The New Public Health* (Open UP: Milton Keyes).

Bardo, J. and Bardo, D. (1983), 'A re-examination of subjective components of community satisfaction in a British new town', *Journal of Social Psychology* 120: 35–43.

Ben Shlomo, Y. et al. (1996), 'Does the variation in socioeconomic characteristics of an area affect mortality?', *BMJ* 312: 1013–1014.

Blomley, N. (1994), 'Health, geography and society', in M.V. Hayes, L.T. Foster and H.D. Foster (eds) *The Determinants of Population Health: A Critical Assessment. Western Geographical Series, vol. 29.* (Victoria: University of Victoria).

Brown, B., Perkins, D.D. and Brown, G. (2003), 'Place attachment in a revitalizing neighborhood: individual and block levels of analysis', *Journal of Environmental Psychology* 23: 259–271.

Buckner, J. (1988), 'The development of an instrument to measure neighborhood cohesion', *American Journal of Community Psychology* 16: 771–791.

Butz, D. and Eyles, J. (1997), 'Reconceptualizing senses of place: social relations, ideology and ecology', *Geografiska Annaler* 79B: 1–25.

Chavis, D. and Pretty, G. (1999), 'Sense of Community. Advances in measurement and Application', *Journal of Community Psychology* 27: 635–642.

Cheng, A.S., Kruger, L.E. and Daniels, S.E. (2003), "Place' as an integrating concept in natural resource politics: propositions for a social science research agenda', *Society and Natural Resources* 16: 87–104.

Clark, J. and Stein, T.V. (2003), 'Incorporating the natural landscape within an assessment of community attachment', *Forest Science* 49(6): 868–876.

Collins, J.W. Jr., David, R.J., Symons, R., Handler, A., Wall, S. and Andes, S. (1998), 'African-American mothers' perception of their residential environment, stressful life events, and very low birth weight', *Epidemiology* 9(3): 286–289.

Cresswell, T. (1996), *In Place/Out of Place: Geography, Ideology and Transgression* (Minneapolis: University of Minnesota Press).

Cuba, L. and Hummon, D.M. (1993), 'A place to call home: identification with dwelling, community, and region', *The Sociological Quarterly* 34(1): 111–131.

Dunn, J.R. and Hayes, M.V. (2000), 'Social inequality, population health, and housing: a study of two Vancouver neighborhoods', *Social Science and Medicine* 51: 563–587.

Ellaway, A. and Macintyre, S. (2001), 'Women in place: gender and perception of neighborhoods and health in the West of Scotland', in I. Dyck, N.D. Lewis and S.L. McLafferty (eds.) *Geographies of Women's Health* (London: Routledge).

Ewart, C.K. and Suchday, S. (2002), 'Discovering how urban poverty and violence affect health: development and validation of a neighborhood stress index', *Health Psychology* 21(3): 254–262.

Eyles, J. (1985), *Senses of Place* (Cheshire: Silverbrook Press).

—— (1999), 'Health, environmental assessment and population health', *Canadian Journal of Public Health* 90: S31–S34.

Frank, J. (1995), 'The determinants of health', *Current Issues in Public Health* 1: 233–240.

Fuller, T.D., Edwards, J.N., Sermsri, S. and Vorakitphokatorn, S. (1993), 'Housing, stress, and physical well-being: evidence from Thailand', *Social Science and Medicine* 36(11): 1417–1428.

Galster, G. (2001), 'On the nature of neighbourhood', *Urban Studies* 38(12): 2110–2124.

Gatrell, A. (2002), *Geographies of Health* (Oxford: Blackwell).

Gesler, W. (1993), 'Therapeutic landscapes: theory and a case study of Epidauros, Greece', *Environment and Planning D: Society and Space* 11: 171–189.

—— (1992), 'Therapeutic landscapes: medical issues in light of the New Cultural Geography', *Social Science and Medicine* 34: 735–746.

Giddens, A. (1985), 'Time, space and regionalization', in D. Gregory and J. Urry (eds) *Social Relations and Spatial Structures* (New York: St. Martin's Press).

Glynn, T. (1981), 'Psychological sense of community. Measurement and application', *Human Relations* 34: 780–818.

Goudy, W.J. (1982), 'Further Considerations of Indicators of Community Attachment', *Social Indicators Research* 11: 181–192.

Hamilton, N. and Bhattia, T. (1996), *Population health promotion* (Ottawa, Health Canada).

Hay, R. (1998), 'A rooted sense of place in cross-cultural perspective', *The Canadian Geographer* 42(3): 245–266.

Hayes, M.V. and Dunn, J.R. (1998), *Population Health in Canada: A Systematic Review* (Ottawa: Canadian Policy Research Networks).

Healey, P. (1998), 'Institutionalist theory, social exclusion and governance', in A. Madanipour, G. Cars and J. Allen (eds) *Social Exclusion in European Cities*. (London: Jessica Kingsley Publishers).

Heymann, J. et al. (eds) (2006), *Healthier Societies* (New York: Oxford UP).

Hummon, D.M. (1990), *Commonplaces: Community Ideology and Identity in American Culture* (Albany: State University of New York Press).

James, J. and Eyles, J. (1999), 'Perceiving and representing both health and the environment: an exploratory investigation', *Qualitative Health Research* 9(1): 86–104.

Johnson, R.B. and Christensen, L.B. (2000), *Educational Research: Quantitative and Qualitative Approaches* (Boston: Allyn and Bacon).

Jorgensen, B.S. and Stedman, R.C. (2001), 'Sense of place as an attitude: lakeshore owner attitudes toward their properties', *Journal of Environmental Psychology* 21: 233–248.

Judge, K. (1995), 'Income distribution and life expectancy', *BMJ* 311: 1282–1285.

Kaplan, G. and Lynch, J. (1997), Editorial: 'Whither studies on the socioeconomic foundations of population health?', *American Journal of Public Health* 87: 1409–1411.

Kearns, R.A. (1995), 'Medical geography: making space for difference', *Progress in Human Geography* 19: 251–259.

—— (1991), 'The place of health in the health of place: the case of the Hokianga Special Medical Area', *Social Science and Medicine* 33: 519–530.

Kearns, A., Hiscock, R., Ellaway, A. and Macintyre, S. (2000), 'Beyond four walls. The psycho-social benefits of home: evidence from west central Scotland', *Housing Studies* 15: 387–410.

Labonte, R. and Schrecker, T. (2007), 'Globalization and the social determinants of health', *Globalization and Health* 3(5), 3(6), 3(7).

Lalli, M. (1992), 'Urban-related identity: theory, measurement, and empirical findings', *Journal of Environmental Psychology* 12: 285–303.

Liaschenko, J. (1994), 'The moral geography of home care', *Advances in Nursing Science* 17: 16–26.

Lillie-Blanton, M. and LaVeist, T. (1996), 'Race/ethnicity, the social environment and health', *Social Science and Medicine* 43: 83–91.

Luginaah, I.N., Taylor, S.M., Elliott, S.J. and Eyles, J.D. (2002), 'Community reappraisal of the perceived health effects of a petroleum refinery', *Social Science and Medicine* 55: 47–61.

Lynch, J. and Kaplan, G. (1997), 'Understanding how inequality in the distribution of income affects health', *Journal of Health Psychology* 2: 297–314.

Macintyre, S., Maciver, S. and Sooman, A. (1993), 'Area, class and health: should we be focusing on places or people?', *Journal of Social Policy* 22: 213–234.

Macintyre, S. (1997), 'The black report and beyond: what are the issues?', *Social Science and Medicine* 44: 723–745.

Marcus, A.C. and Crane, L.A. (1986), 'Telephone surveys in public Health Research', *Medical Care* 24: 97–112.

McDowell, L. (1997), *Undoing Place? A Geographical Reader* (London: Arnold).

McKeown, T. (1976), *The Role of Medicine* (Oxford: Blackwell).

Meegan, R. and Mitchell, A. (2001), '"It's not community round here, it's neighbourhood": Neighbourhood change and cohesion in urban regeneration policies', *Urban Studies*, 38(12): 2167–2194.

Mellor, J. and Milyo, J. (2001), 'Re-examining the evidence of an ecological association between income inequality and health', *Journal of Policy, Politics and Law* 26: 487–522.

Mesch, G.S. and Manor, O. (1998), 'Social ties, environmental perception, and local attachment', *Environment and Behavior* 30(4): 504–519.

Mitchell, D. (2003), 'Cultural landscapes: just landscapes or landscapes of justice?', *Progress in Human Geography* 27(6): 787–96.

Muhajarine, N., Labonte, R., Williams, A. et al. (2003), 'Person, perception, and place: what matters to health and quality of life' (submitted to *Social Science and Medicine*).

Muntaner, C. (2004), 'Social Capital, social class and the slow progress of psychosocial epidemiology', *International Journal of Epidemiology* 33: 674–680.

Mustard, F. and Frank, J. (1994), 'The determinants of health', in M. Hayes et al. (eds), *The Determinants of Population Health*, Western Geographical Series 29: University of Victoria: 7–48.

Nasar, J. and Julian, D. (1995), 'The psychological sense of community in the neighborhood', *Journal of the American Planning Association* 61: 178–184.

Oakes, T. (1997), 'Place and the paradox of modernity', *Annals of the Association of American Geographers* 87: 509–531.

Peter, E. (2002), 'The history of nursing in the home', *Nursing Inquiry* 9: 65–72.

PHAC (Public Health Agency of Canada) (2002), Using a population health approach to promote health, www.phac-aspc.gc.ca/ph-sp/phdd/work/index.html#approach (accessed 6 Dec, 2007).

Pretty, G.H., Chipuer, H.M. and Bramston, P. (2003), 'Sense of place amongst adolescents and adults in two rural Australian towns: the discriminating features of place attachment, sense of community and place dependence in relation to place identity', *Journal of Environmental Psychology* 23: 273–87.

Relph, E. (1976), *Place and Placelessness* (London: Pion).

Ryan, G.W. and Bernard, H.R. (2000), 'Data management and analysis methods', in N.K. Denzin and Y.S. Lincoln (eds) *Handbook of Qualitative Research.* Second Edition. (Thousand Oaks: Sage Publications).

Sampson, R.J. (1988), 'Local friendship ties and community attachment in mass society: a multilevel systemic model', *American Sociological Review* 53: 766–779.

Shye, S. (ed.) (1978), *Theory Construction and Data Analysis in the Behavioral Sciences* (San Francisco: Jossey-Bass Publishers).

Smith, M.P. (1979), *The City and Social Theory* (Oxford: Basil Blackwell).

Stangor, C. (1998), *Research Methods for the Behavioral Sciences* (Boston: Houghton Mifflin Company).

Stedman, R.C. (2003a), 'Sense of place and forest science: toward a program of quantitative research', *Forest Science* 49(6): 822–829.

—— (2003b), 'Is it really just a social construction? The contribution of the physical environment to sense of place', *Society and Natural Resources* 16: 671–685.

Subramanian, S. et al. (2001), 'Does the state you live in make a difference?' *Social Science and Medicine* 53: 9–19.

Syme, S.L. (1994), 'The social environment and health', *Daedalus* 123: 79–86.

Tashakkori, A. and Teddlie, C. (eds) (2003), *Handbook of Mixed Methods in Social and Behavioral Research* (Thousand Oaks: Sage Publications Inc).

Theodori, G.L. (2001), 'Examining the effects of community satisfaction and attachment on individual well-being', *Rural Sociology* 66(4): 618–628.

Tuan, Y. (1974), *Topophilia: A Study of Environmental Perception, Attitudes and Values* (New Jersey: Prentice-Hall).

Turrell, G. et al. (2007), 'Do places affect the probability of death in Australia?', *Journal of Epidemiology and Community Health* 61: 13–19.

UN (United Nations) (2005), Millennium development goals, www.un.org/millenniumgoals (accessed 6 December 2007).

Vega, J. et al. (2006), 'The commission on the social determinants of health', *Plos Med*: 3(6): e106.

Wakefield, S.E.L., Elliot, S.J., Cole, D.C. and Eyles, J.D. (2001), 'Environmental risk and (re)action: air quality, health, and civic involvement in an urban industrial neighbourhood', *Health and Place* 7: 163–177.

Warin, M., Baum, F., Kalucy, E., Murray, C. and Veale, B. (2000), 'The power of place: space and time in women's and community health centres in South Australia', *Social Science and Medicine* 50: 1863–1875.

Watts, M. (2000), *Community. The Dictionary of Human Geography* (Oxford: Blackwell Publishers).

WHO (World Health Organization) (2007), *Achieving Health Equity* (Geneva, WHO).

Wilkinson, R. and Marmot, M. (2003), *Social Determinants of Health: The Solid Facts* (Copenhagen: WHO).

Williams, A. (1999), *Therapeutic Landscapes: The Dynamic Between Place and Wellness* (Maryland: University Press of America).

—— (1998), 'Therapeutic landscapes in holistic medicine', *Social Science and Medicine* 46: 1193–1203.

Williams, A. and Cutchin, M. (2002), 'The rural context of health care provision', *Journal of Interprofessional Care* 6(2): 107–115.

Williams, A., Abonyi, S., Dunning, H., Carr, T., Holden, B., Labonte, R., Muhajarine, N. and Randall, J. (2002), *Quality of Life in Saskatoon, SK: Achieving a Healthy, Sustainable Community: Research Summary*. (Community-University Institute for Social Research (CUISR): University of Saskatchewan, Saskatoon, Saskatchewan).

Williams, D.R. and Vaske, J.J. (2003), 'The measurement of place attachment: validity and generalizability of a psychometric approach', *Forest Science* 49(6): 830–840.

Yung, L., Freimund, W.A. and Belsky, J.M. (2003), 'The politics of place: understanding meaning, common ground, and political difference on the Rocky Mountain Front', *Forest Science* 49(6): 855–866.

Chapter 2

A Sense of Place, A Sense of Well-being

Lily DeMiglio and Allison Williams

> Whatever a guidebook says, whether or not you leave somewhere with a sense of place is entirely a matter of smell or instinct. There are places I've been which are lost to me (146). Once *in* a place, that journey to the far interior of the psyche begins or it doesn't. *Something* must make it yours, that ineffable something no book can capture (148).
>
> Frances Mayes (1997)

This chapter provides a review of the literature on sense of place, in an effort to enhance communication about sense of place and to recognize its unique contribution as a place-related construct to health. For the purpose of this chapter, health will be broadly defined in accordance with the World Health Organization (WHO) definition[1] and as such, *health* and *well-being* will be used interchangeably. The overarching aim is to frame the sense of place literature within the context of health. To begin with, an overview of the sense of place concept will be provided to highlight its use in a wide range of disciplines, including architecture, forestry, geography and psychology. This objective involved consulting and retrieving a myriad of sources, including: books, abstracts, citations and journal articles, both empirical and theoretical, from a variety of geography, psychology and sociology databases.[2] The results yielded a collection of over a thousand abstracts. The sources included in the present discussion represent those works that most encompass the concept of

1 The WHO (1946) defines health as "a state of complete physical, mental and social well-being and not merely the absence of disease or infirmity".

2 Sense of place and seventeen related terms were employed as search terms across several online databases and search engines. The seventeen search terms were: community satisfaction, (psychological) sense of community, sense of community, community identity, neighbourhood satisfaction, neighbourhood cohesion, neighbourhood attachment, neighbourhood identity, rootedness, meaningfulness of place, place attachment, place identity, place dependence, ecological attachment, belonging, local sentiment and topophilia. The online databases and search engines included: Applied Social Sciences Index and Abstracts (ASSIA), Full SAGE text collection (1969-present), GEOBASE, Health and Psychosocial Instruments (HAPI), Historical abstracts, Institute for Scientific Info (ISI) Journal Citation Reports (2002–present), Journal Citation Reports (JCR) Social Science Edition, JSTOR, PAIS (Scholars Portal International, social sciences), Psychology: Find Articles (1998–present), PsycINFO, Scholars Portal, Science Direct, Social Sciences Abstracts, Social Science Citation Index, Sociological Abstracts, Urban Studies Abstracts, Urban Studies and Planning: A Sage Full Text Collection and Web of Science.

sense of place relative to the objective of the discussion. In the second section of the chapter, the concept of *place* will be discussed to help delineate sense of place as the product of the relationship between people and places. The final section will explore a number of studies that provide evidence to support the connection between sense of place and well-being in addition to the variables that influence sense of place. This section demonstrates that the relationship between sense of place and well-being has yet to be fully investigated and warrants further exploration in light of present day human and environmental health challenges. For practical reasons, the following review is non-exhaustive. Studies, which focused on sense of place with no apparent connection to well-being, were not included in the literature review given both the objective and the limited scope of the discussion.

The Conceptualization of Sense of Place: An Overview

Consulting the works of the scholar responsible for developing the term *sense of place* is not possible since it has been widely theorized by more than one particular person (e.g., Relph 1976; Tuan 1980; Steele 1981; Eyles 1985; Jackson 1994 and Hay, 1998). The term *genius loci*[3] (i.e. spirit of place) is sometimes cited as the precursor to sense of place. According to the Romans, genius loci meant that places were safeguarded by spirits (Relph 1976; Lewis 1979 and Jackson 1994). Citizens and visitors paid homage to these guardian spirits through ceremonial celebrations but by the 1700s, the idea of spirits safeguarding places was no longer part of the interpretation of the term[4] (Jackson, 1994). Edward Relph (2006) concurs that "spirit of place, the inherent properties that lend identity to somewhere, can be distinguished from sense of place,[5] – the faculty by which that identity is perceived" (19). Yi-Fu Tuan (1996) adds, "place may be said to have 'spirit' or 'personality', but only human beings can have a sense of place" (446). According to landscape writer, J.B. Jackson (1994), although many do not directly associate sense of place with spirits, as a modern day equivalent of genius loci, a spiritual undertone remains embedded in the concept:

> We now use the current version to describe the *atmosphere* to a place, the quality of its *environment*. Nevertheless, we recognize that certain localities have an attraction which gives us a certain indefinable sense of well-being and which we want to return to, time and again. So that original notion of ritual, of repeated celebration or reverence, is still inherent in the phrase. It is not a temporary response, for it persists and brings us back, reminding us of previous visits (158).

3 The terms *sense of place*, *genius loci*, *character*, and *appearance*, have been used interchangeably in the field of urban planning and design according to Jiven and Larkham (2003). However, they do note that architect Norberg-Schulz (1980) advised against using *sense of place* and *genius loci* interchangeably.

4 It should be noted that the notion of spirits safeguarding places still exists in some cultures today (e.g. Aboriginal peoples).

5 Relph refers to this concept as "sense of *a* place" in Chapter 3 as opposed to "sense of place".

Thus, according to Jackson (1994), sense of place has the ability to impart a sense of well-being. In order to further examine the relationship between sense of place and well-being, it is important to first synthesize the seminal contributions that have played a role in the evolution of the sense of place construct.

One of the earliest references to sense of place dates back to a 1965 report prepared by a group of American geographers, summoned by the chairman of the Earth Sciences Division of the National Academy of Sciences – National Research Council, to evaluate the prospect of the discipline of geography as a scientific research field. A sense of place, according to the committee, is "a compound of a sense of 'territoriality,' physical direction, and distance, [that] is very deeply ingrained in the human race" (1965, 7). Subsequently, under the section titled, *A Place for the Unusual Idea*, the committee acknowledged the need for more research about the concept. The following excerpt suggests that the committee's view of sense of place is from a neurological or biological perspective. As such, *sense* of place is represented as a faculty of perception and it is presumed to be biologically-based, similar to the five human senses:

> … little is known as yet about what we earlier called the "sense of place" in man [sic]. Its secrets are still locked from us in our inadequate knowledge of nervous systems. Someday, when the study of nervous systems has advanced sufficiently, a startling and perhaps revolutionary new input may reach geographical study in a full descriptive analysis of the sense of place. We hope that if a geographer has an interesting opportunity with the proper collaboration to delve into the mysteries of the sense of place, he may somewhere find a sympathetic ear among those who have funding responsibilities (67–68).

The committee's interpretation of sense of place does indeed seem plausible. Therefore, just as the five senses play a role in sustaining health and well-being, it is possible that sense of place does the same. As a method of perception, sense of place is comparable to the five human senses in that it is often difficult to communicate or explain. In terms of the sense of taste, while one may consider a food to taste delicious, someone else may consider the same food to be unpleasant. Furthermore, one may consider a food to taste delicious but finds it difficult to describe the delicious taste. The same holds true for sense of place perception. Although we understand what is meant by sense of place, it remains a difficult concept to express (Beatley 2003).

Buttimer (1980) recalls her sense of place of Ireland based on her early-life experiences and affirms that indeed, sense of place is difficult to describe; her sense of place is perceived using a combination of the five human senses:

> … I recall the feel of grass on bare feet, the smells and sounds of various seasons, the places and times I meet friends on walks, the daily ebb and flow of milking time, meals, reading and thinking, sleeping and walking. Most of this experience is not consciously processed through my head – that is why words are so hard to find – for this place allows head and heart, body and spirit, imagination and will to become harmonized and creative (172–173).

Buttimer's description corresponds to urban planner Kevin Lynch's (1976) concept of the *sensed quality of a place*, this is the overall perception of a place based on the summation of separate sensations. He maintains that the sensed quality of a place has an effect on well-being since many physical processes (e.g. breathing and hearing) are mediated by sensory cues. According to Lynch (1976), the sensed quality of a place encompasses:

> … what one can see, how if feels underfoot, the smell of the air, the sounds of bells and motorcycles, how patterns of these sensations make up the quality of places, and how that quality affects our immediate well-being, our actions, our feelings, and our understandings … What is sensed has fundamental and pervasive effects on well-being (8–9).

For this reason, he proposes several guidelines for planners to employ to help improve the sensory experience of an environment (e.g. controlling noise pollution). Howard Frumkin (2003) also acknowledges that the characteristics of places have the ability to influence health. He recognizes sense of place as a potential contributing factor to well-being within the field of public health; he wrote, "if sense of place has benefits for health and well-being, then understanding how to design for it may have real public health value" (1452). Although Frumkin proposes sense of place as a public health construct, he acknowledges the fact that the concept remains loosely defined.

It is possible for individuals to share a common sense of place (i.e. for the same physical place) or for it to be an individually distinctive construction (Pred 1983). In the same regard, it is possible for individuals to share the same interpretation for the words *sense* and *place* or for their interpretations to be entirely different. In terms of semantics, the words *sense* and *place* themselves have multiple meanings that might in turn influence the understanding of *sense of place*. In accordance, Williams (2007) offers three different viewpoints of sense of place, the first of which parallels the interpretation offered by the National Academy of Sciences:

1. sense of place as a faculty or capacity (e.g. a keen sense of place similar to a keen sense of humour or a keen sense of smell)
2. sense of place as cognitions of place (e.g. knowledge and awareness of place)
3. sense of place as the character of a place (e.g. atmosphere of place)

(Williams 2007)

In the forty-some years that have passed since the publication of the Earth Sciences Division of the National Academy of Sciences – National Research Council report, many have attempted to develop the concept of sense of place from their own perspectives and research is currently underway to determine the location of the sense of place in the brain (Bond 2006). The Dictionary of Human Geography (2000) recognizes that although sense of place is rooted in geography, it is a multifaceted construct:

> Originating in studies of the physical characteristics and qualities of geographical locations as appropriated in human experience and imagination, sense of place has increasingly been examined in human geography as an outcome of interconnected psychoanalytic,

> social and environmental processes, creating and manipulating quite flexible relations with physical place (Cosgrove 2000, 731).

From this excerpt, it is evident that the field of human geography acknowledges that sense of place is the product of the relationship that individuals form with places. For this reason, it is not surprising that the sense of place construct caught the attention of researchers outside the field of geography. Over the past few decades, sense of place has been explored through the lens of numerous disciplines, including: architects (e.g. Ouf 2001; Jiven and Larkham 2003), psychologists (e.g., Steele 1981; Hay 1998), sociologists (Hummon 1992), anthropologists (e.g. Altman and Low 1992[6]), environmental scientists (e.g. Williams and Stewart 1998; Stokowski 2003), travel writers (e.g. Mayes 1997) and some have even attempted to measure the concept (e.g. Shamai and Kellerman 1985; Shamai 1991; Jorgensen and Stedman 2001; Stedman 2003; Shamai and Ilatov 2004; Jorgensen and Stedman 2006). Regardless of the discipline or methodological approach, it is evident that research on sense of place stems from a vested interest in understanding the complex forces that shape the relationship between people and places.

Sense of Place: A Multidisciplinary Construct

The wide appeal of sense of place may inadvertently have caused Peirce Lewis (1979) to question the concept in the late 1970s: "Does it really exist, apart from a rather trendy phrase that seems to be on everyone's lips these days?" (26). His reservations may have based on the fact that *sense of place* is often used as a catchphrase (Jackson 1994) and informally in everyday conversation. Notwithstanding, the value, utility and relevance of sense of place has not waned; this is evidenced by its ever-increasing use across disciplines. Given its versatility as a construct, it is apparent that sense of place has been interpreted in many ways. More often than not, interpretations of sense of place are discipline-specific and they will be further discussed herein. The following collection of excerpts about sense of place is multidisciplinary. Yet the works suggest that there is a connection between sense of place and well-being but this relationship is not an obvious one.

Some consider sense of place to contribute to well-being more so than others. Despite his uncertainty, Lewis (1979) went on to endorse sense of place as an essential for life and as a need that is necessary to fulfill in order to sustain personal well-being:

> To have a sense of place – to sense the spirit of place – one's own place – is as indispensable to the human experience as our basic urges for food, or for sex. Indeed, we even use similar words to describe our emotions. We speak of "hunger" to return home, just as we speak of hungering for food. Our affection for certain beloved places is as intense, perhaps as our affection for people we love. To be sure, one can overindulge one's hunger for food: that is gluttony. One can overindulge one's sexual urge, and that is debauchery. So also, one can overindulge a love of place, and more than a few wars have been fought

6 Patterson and Williams (2005) note that *sense of place* and *place attachment* are often used interchangeably as is the case in Altman and Low's (1992) *Place Attachment*.

> "in defense of homeland, or fatherland," or what have you, and they were murderous wars indeed. But I do not think that one can survive as a humane creature on this earth without special attachments to special places (29).

Years earlier, Ian Nairn (1965), an architectural critic, expressed a similar viewpoint:

> I am not setting up as a psychologist, but it seems a commonplace that almost everyone is born with the need for identification with his surroundings and a relationship to them – with the need to be in a recognizable place. So sense of place is not a fine-art extra, it is something which man [sic] cannot afford to do without (6).

In his book *The Sense of Place*, organizational and environmental psychologist Fritz Steele (1981), did not go as far as Nairn (1965) and Lewis (1979) to suggest that sense of place is a human need, however, he did propose that sense of place is related to emotional well-being. As such, sense of place is said to influence a person's mental state (i.e. both conscious and unconscious) and thus it will affect how a person will respond to a place, which is conveyed through their actions and/or emotions.

> The *sense of place*, … is the particular experience of a person in a particular setting (feeling stimulated, excited, joyous, expansive, and so forth) … [it] is the pattern of reactions that a setting stimulates for a person. These reactions are a product of both features of the setting and aspects the person brings to it … [it] is an interactional concept: a person comes into contact with a setting, which produces reactions. These include feelings, perceptions, behaviours, and outcomes associated with one's being in that location. Sense of place is not limited just to the experience of which the person is consciously aware; it includes unnoticed influences, such as a consistent avoidance of doing certain things in that particular place (Steele 1981, 11–12).

So it seems from Steele's depiction that sense of place has the capability to not only influence one's psychological or emotional well-being but also one's actions. Sociologist, David Hummon (1992) took Steele's conception of sense of place further by adding, "sense of place is inevitably dual in nature, involving both an interpretive perspective *on* the environment and an emotional reaction *to* the environment" (262). The act of *interpretation* proposed by Hummon coincides with Tuan's (1977) notion that indiscriminate space turns into place "as we get to know it better and endow it with value" (6). Architect, Christian Norberg-Schulz (1969) illustrates Tuan's point of view by emphasizing the importance of imbuing places with meaning: he wrote, "when we are traveling in a foreign country, space is 'neutral', that is, not yet connected with joys and sorrows. Only when it becomes *a system of meaningful places*, does it become alive to us" (224). It should be noted that unlike Steele (1981), Tuan (1980) affirms that sense of place is a completely conscious experience and he differentiates it from another concept, *rootedness*, which is an unconscious experience and thus a more deeply ingrained phenomenon. Therefore, there is a component of awareness attached to sense of place, and according to Tuan (1980), sense of place can be "achieved and maintained", which is not the case for rootedness (4). Moreover, Relph (2006) states that sense of place is a dynamic construct such that it tends to vary as a function of time and culture. Forestry scientists Williams and Stewart (1998) realized the significance of sense of place and synthesized the work of scholars from a variety of fields to better define the concept. Based on their comprehensive definition, sense of place is an umbrella concept that captures the essence of the relationship people form with places; as such, it encompasses the following elements:

1. the emotional bonds that people form with places (at various geographic scales) over time and with familiarity with those places;
2. the strongly felt values, meanings, and symbols that are hard to identify or know (and hard to quantify), especially if one is an "outsider" or unfamiliar with place;
3. the valued qualities of a place that even an "insider" may not be consciously aware of until they are threatened or lost;
4. the set of place meanings that are actively and continuously constructed and reconstructed within individual minds, shared cultures, and social practices; and
5. the awareness of the cultural, historical, and spatial context within which meanings, values, and social interactions are formed

(Williams et al. 1998, 19)

This detailed conceptualization, together with those already presented, consistently portrays sense of place as the product of the relationship between people and places. Individuals establish relationships with a variety of different places throughout their lives and the quality or strength of these relationships can vary (Manzo 2003). Sense of place can be positive or negative, weak or strong (Relph 1976; Eyles 1985 and Manzo 2003). Furthermore, it should be noted that sense of place varies depending on a number of factors, such as the nature of the place itself and time; these and other factors will be further examined.

Sense of Place: The Dimension of Place as a Material Entity

Prior to presenting evidence to substantiate the relationship between sense of place and well-being as well as discussing variables that influence sense of place, the component of *place* will be further explored. Unlike other potentially place-related terms (e.g. social capital, social cohesion), sense of place is exclusively place-based; for this reason, the properties of the physical environment contribute to it. Jackson (1994) acknowledged that sense of place is largely influenced by the "quality of [the] *environment*" (158). Hence, sense of place is interpreted as a formulated response towards the characteristics or aspects of the environment (i.e. natural, built and social). Therefore, the type of relationship that one forms with the environment is an individual process and it may depend on the context of the place itself. For instance, Butz and Eyles (1997) introduced the concept of "ecological" senses of place, which is "the knowledge of a place's ecological characteristics that yield meanings which make persons identify with place" (10). More precisely, ecological senses of place result from the interaction between people and components of their environment; they are shaped but not dictated by the characteristics of the environment alone.

Living in a global world has greatly increased the number of places people encounter throughout their lifetime. As more people travel between places (e.g. with the ease of commuting), there is said to be a weakening effect on sense of place. Relph (1976) stated, "... for the primitive hunter or medieval artisan a sense of belonging to his place imbued his whole existence, for the modern-city dweller it is rarely in the foreground and can usually be traded for a nicer home in a better neighbourhood" (66). Although subsequent research suggests that city dwellers have the capacity to develop a sense of place and belonging, it is evident that several dimensions of modern life now challenge sense of place. In his book *No Sense of Place*, sociologist, Meyrowitz (1985) reveals that media, particularly television, has blurred the boundaries of place. Butz and Eyles (1997) offer, "it may be that our strongest senses of place are of places, which have no physical environmental grounding (for example, imagined places or 'cyberplace'), and these can hardly be described as ecologically grounded" (11). However, it terms of cyberplace or cyberspace (i.e. the Internet), the question remains whether it is classifiable as a *place* or whether it is a medium used to escape place. It can be argued that websites such as "Facebook" and "MySpace", in addition to online chat groups, provide their users with a sense of community, and thus a sense of place. Although it seems clear that globalization and new technologies (e.g. the Internet) will surely influence the future of sense of place, however, more research is necessary to determine the capacity of its effect, on both sense of place and well-being.

It is necessary to recognize that many different senses of place exist, given the inter-subjective nature of the concept. Relph (1976) recognizes that everyone will experience sense of place but the degree will vary from person to person. He describes two forms of sense of place, "authentic and genuine" and "inauthentic and contrived or artificial" (1976, 63). As such, an authentic sense of place is experienced by individuals who achieve a sense of belonging to place (e.g. their home, community or country) and thereby contributes to an individual's identity. In contrast, an inauthentic sense of place results from the inability to develop a meaningful relationship with the environment.

Hummon (1992) discusses different types of senses of place that exist among citizens of the same community; his inventory includes degrees of "rootedness", "alienation", "relativity" and "placelessness" (263). Relph (1976) was first to describe "placelessness" (i.e. when there is no sense of place) (79), when places are stripped of their unique attributes and commonalities between places start to exist, thus compromising place identity. According to Hummon, different "community sentiments" resulting from how people think and feel about their community will evoke a variety of senses of place. He describes various degrees of rootedness but at its simplest, a rooted sense of place is feeling an emotional connection to the place where one resides (i.e. both your home and vicinity around your home).

Hummon also describes place alienation, which results from, for instance, the inability to fully experience place due to immobility or the inability to establish or develop a sense of belonging (i.e. feeling "displaced") (269). In his description of alienation, Hummon offers the example of a woman who still feels a deep connection towards the community where she was raised and fails to develop a sense of belonging to her new/current community more than ten years after moving. In contrast, his notion of placelessness involves a total lack of emotional ties to the community or identification with the community. Moreover, although the concept of place relativity is associated with the lack of feelings towards community, it also involves experiencing a sense of home regardless of the place where it exists. For instance, Hummon refers to the experience of a man who had lived in a number of different communities; the man revealed that each place became "home" for him after a period of adjustment. From this example, it is clear that sense of place may vary over time. Variables that influence sense of place will be further discussed in the following section.

Factors Influencing the Relationship between Sense of Place and Well-being

A number of factors have been highlighted as playing an influential role in the relationship between people and place; these include: time, residential status, age, ethnicity and the characteristics of the place itself. These variables mediate the relationship that people form with places and in turn, the sense of well-being derived from sense of place.

The relationship between sense of place and time is discussed in the work of Taylor and Townsend (1976). Through the administration of a survey, they explored sense of place in four towns located in North-East England. It was found that elements of the environment (i.e. both natural and built) were not fully related to how individuals view place. Instead, it was the extent of experience (i.e. length of residence) that an individual had with the particular place that, in turn, influenced the perception of place; those individuals over the age of 65 expressed a stronger sense of place.

Through processes such as immigration, relocation and displacement, individuals are often required to rebuild their relationship with place in order to establish a sense of place. Mazumdar, Mazumdar, Docuyanan and McLaughlin (2000) explored Vietnamese immigrants living in an ethnic enclave in the United States of America and their relationship to place; they found architectural design, socialization and the celebration of rituals to contribute to sense of place. Thus, sense of place encompasses aspects of both the social and physical environments. The immigrants established a sense of place by incorporating architectural features reminiscent of their homeland in the design of buildings and landscaping within the enclave; the Vietnamese immigrants attempted to build a positive sense of place in their new community with elements of the past. Although this study shows the overall experience of immigrants to be positive, it is not the case for all immigrants.

Ortiz, Garcia-Ramon and Prats (2004) uncovered weak senses of place in their study, which explored three different groups of women (i.e. non-immigrants, European immigrants and non-European immigrants) through the use of common public space in a neighbourhood in Barcelona, Spain. Sense of place differed among the three groups of women yet it was unclear whether the use of public space had a direct effect on women's sense of place. The local women experienced a decline in their sense of belonging in part due to perceptions of neighbourhood change influenced partly as a result of an influx of immigrants into the area. Both of the immigrant groups were still in the process of developing a sense of place. It is possible that the negative sentiments of the local native women towards the immigrant women could evolve into what Relph (1997) refers to as a "poisoned" sense of place, which involves "an excess of local or national zeal … in that other places and peoples are treated with contempt … sense of place carries within itself a blindness and a tendency to become a platform for ethnic nationalist supremacy and xenophobia" (222).

Similar to Ortiz and colleagues (2004), Billig (2005) also discusses sense of place from the perspective of women. More specifically, she investigates the sense of place of women living in newly built housing developments (i.e. found within or near old neighbourhoods) in contrast to those living in older developments in Israel. She asserts that the concept of sense of place might play a role in urban planning initiatives. The study was conducted by using qualitative methods whereby interview data was collected to evaluate for sense of place variables. In sum eight variables[7] were categorized as either "variables of behavior" or "variables of personal feeling/affective variables" (121). For example, a variable that was grouped in the behavior category was the use of public space and amenities while the sentiment of belonging was included under the rubric of personal feeling variables. The results show that senses of place differ in diverse neighbourhoods, whether new or old, due to, amongst other factors, the heterogeneity of residents (e.g. socioeconomic and life style differences), varying degrees of social ties among residents as well as varying degrees of interaction with aspects of the built-environment.

Other studies show that sense of place is less a matter of residential status and more closely associated with the significance that people attach to place. For instance, Kianicka, Buchecker, Hunziker and Muller-Boker (2006) examine sense of place differences in local citizens and visitors to a Swiss Alpine village. Their findings show that the components that influenced the two groups' sense of place were the same but that the level of significance attached to each component varied between groups. For local citizens, sense of place was most influenced by personal aspects such as social networks, their homes and their personal histories associated with place. For tourists (or those with less experience in a place) sense of place was most influenced by dimensions of the natural environment that contribute to their recreational experience of place.

Sense of place can also be influenced by the type of relationship that people form with place and its attributes. More specifically, the type or degree of sense of place is often shaped by what the place has to offer. These place characteristics or variables are construed as having the capability to influence a person's well-being.

---

7 The variables of behavior included: differences in norms and life styles between population groups, characteristics of relations among the same population group, characteristics of relations with other populations group(s) and use of public space and of public services (Billig 2005, 121). The variables of personal feeling included: satisfaction with or aversion to the physical environment, feeling of belonging to the physical environment, feeling of belonging to a community and feeling of security in public space (121).

In these instances, the relationship that individuals establish with place is often based on whether or not the place offers amenities and opportunities that uphold or improve their standard of living. Eyles (1985) examined the attitudes of residents of Towcester, a small town in Northhamptonshire, England and categorized ten different senses of place.[8] In doing so he offered the following disclaimer:

> It will be noted that convenient but sometimes clumsy labels have been used to describe the senses of place … This exercise is not done to suggest that these relationships are in any way categorical or universal. It merely indicates the possibilities of differentiating the sense of place with respect to this particular data-set. Furthermore, it is not suggested that the relative importance of the senses of place can or will be replicated elsewhere. It is just the sense of place categories themselves that may be of wider significance (123).

Although all people experience some form of sense of place, it is obvious that Eyles' comment (1985) clarifies that not all types will be exhibited in a given place. He illustrates that there are many variables that influence sense of place and that these variables are characteristic to the place of interest. Shamai (1991) agrees and adds that most sense of place studies focus on a certain place, which makes it difficult to generalize the conclusions to places, not focused upon in the original study.

Although Eyles' study (1985) is based on a particular town, his sense of place categories are worth noting. The most common categories included "social", "apathetic-acquiescent", "instrumental" and "nostalgic" senses of place. Similar to Relph's (1976) inauthentic sense of place, Eyles (1985) recognized the possibility of being devoid of sense of place and referred to it as "apathetic-acquiescent" (122–124). In contrast, individuals with a social sense of place regard the place where they reside as important because it is a means of facilitating contact with friends and family whereas those with an instrumental sense of place find meaningfulness in what the place offers (e.g. amenities such as shops and employment opportunities). Interestingly, it was noted that there was a negative instrumental sense of place due to the perceived inability of Towcester to provide certain goods and services. Eyles also described a nostalgic sense of place as the product of recalling past sentiments related to place (e.g. memories). Here, it is obvious that the type of sense of place experienced by individuals is truly a reflection of personal views and perceptions. It is also noted that in reflecting about place, some form their opinions based on the present while some recall the past.

Butz and Eyles (1997) revisited the aforementioned study conducted in Towcester, England (Eyles 1985) and compared senses of place exhibited in Towcester to those found in a small agricultural community in Shimshal, Pakistan. Interestingly, an ecological sense of place was found to exist in the Shimshal community but did not appear in Towcester. The two communities did have some common senses of place but they were manifested in different ways mainly due to societal differences. For instance, as an agricultural community, the Shimshal people formed relationships with the natural environment, which most likely contributed to their ecological

---

8 The ten sense of place categories included: social, apathetic-acquiescent, instrumental, nostalgic, commodity, platform/stage, family, way of life, roots and environmental (Eyles 1985, p. 122).

sense of place. Such relationships were not formed in Towcester, as the townspeople livelihood was not heavily reliant on the environment. Thus, senses of place seem to be influenced through the context of community living.

Sense of Place: People, Places and Well-being

Geographers Kearns and Gesler (1998) affirm that Eyles (1985) provides the most comprehensive explanation of the relationship between place and well-being.

> Sense of place, he [Eyles, 1985] proposed, is an interactive relationship between daily experience of a (local) place and perceptions of one's place-in-the-world. This conceptualization sees place as simultaneously center of lived meaning and social position. Place involves an interactive link between social status and material conditions and can be used to interpret a range of situated health effects that imply a link between mind, body, and society (6).

In terms of health outcomes, psychiatrist Fullilove (1996) explains that some psychological disorders emerge when the relationships between people and places disintegrate. In her book, *Root Shock*, Fullilove (2004) explores the emotional turmoil that individuals experience through the process of displacement. On the other hand, mental well-being has been attributed to the maintenance of relationships between people and places established through longevity of residence in a certain place. Williams (2002) introduces the mental health concept, "psychological rootedness" (146), which is related to a strong sense of place and develops through a lengthy experience with place.

According to Wilson (1997), sense of place also plays an important role in early child development processes. She discriminates between places that help to shape a sense of place in children and argues that different characteristics of place help aid in the developmental process. Accordingly, characteristics that are conducive in fostering a sense of place include: a peaceful atmosphere, those settings that invite creative thought and involvement and have the capability of providing memorable experiences. The characteristics of place, Wilson explains, are interpreted by children and help to shape their views about their place in the world. She suggests that certain places contribute to their growth and development and well-being of children and therefore should be maintained and designed accordingly (e.g. school outdoor playgrounds, peaceful refuges). According to Wilson (1997) "attention to creating a sense of place for young children can thus prove helpful in fostering a lifelong commitment to the natural environment" (193).

The notion that sense of place plays a role in the development of environmental stewardship is expanded by Cantrill and Senecah's (2001) concept termed *sense of self-in-place* which explains the perspective through which "humans understand and process various claims and arguments regarding the human relationship to and responsibilities for managing the natural world" (185). Therefore, it is obvious that sense of place plays an important role in not only the development of children but also in the development of the relationship between people and natural environment.

Kaltenborn (1998) examined the relationship between sense of place and individuals' reactions to natural environmental issues in the Norwegian high Arctic. Kaltenborn measured the strength of sense of place (i.e. strong, moderate and weak). He found that those with a strong sense of place were more likely to express concern about environmental impacts and to consider becoming involved in efforts to address issues, in contrast to those with a weak sense of place whom expressed less concern. Here, it is apparent that those with a stronger sense of place exhibited more effort to protect the physical environment than those with a weaker sense of place. Furthermore, the act of protecting place (e.g. from environmental hazards) might be considered a method of protecting self and thus a method of sustaining personal well-being.

Conclusion

This chapter provides an overview of the sense of place construct. Not only does the review examine sense of place as multidisciplinary and complex, but it also examines the relationship between sense of place and well-being. Together, the research studies discussed, show that people establish sense of place with a variety of environments. It is evident that individuals establish relationships with places outside of their home environment. Additionally, the studies cited show that sense of place varies and it is largely based on individual experiences with place and influenced by a number of factors including time, place characteristics and demographic variables such as cultural background, personal history and residential status. Although the relationship between sense of place and well-being is not straightforward, the literature shows that a relationship does indeed exist. In the same regard, the literature review also shows that there is still much to be learned about sense of place, especially whether sense of place is compromised or altered in modern society by globalization and technology. Interdisciplinary research might prove to be a valuable and practical method in order to learn more about this complex yet significant construct.

References

Beatley, T. (2001), 'Home renovations', *Alternatives Journal* 27(2), 32–35.

Billig, M. (2005), 'Sense of place in the neighborhood, in locations of urban revitalization', *GeoJournal* 64, 117–130.

Bond, M. (2006), 'A sense of place', *New Scientist* 189.2541 (March 4, 2006): 50(2).

Buttimer, A. (1980), 'Home, reach, and the sense of place', in A. Buttimer and D. Seamon (eds), *The Human Experience of Space and Place* (London: Croom Helm Ltd.).

Butz, D. and Eyles, J. (1997), 'Reconceptualizing senses of place: Social relations, ideology and ecology', *Geografiska Annaler* 79B(1): 1–25.

Cantrill, J.G. and Senecah, S.L. (2001), 'Using the 'sense of self-in-place' construct in the context of environmental policy-making and landscape planning', *Environmental Science and Policy* 4: 185–203.

Cosgrove, D. (2000), 'Sense of Place', in R.J. Johnston, D. Gregory, G. Pratt and M. Watts (eds), *The Dictionary of Human Geography* (4th ed.) (London: Blackwell Publishers).

Eyles, J. (1985), *Senses of Place* (Warrington: Silverbrook Press).

Frumkin, H. (2003), 'Healthy places: Exploring the evidence', *American Journal of Public Health* 93(9): 1451–1456.

Fullilove, M.T. (1996), 'Psychiatric implications of displacement: contributions from the psychology of place', *The American Journal of Psychiatry* 153(12): 1516–1523.

—— (2004), *Root Shock: How Tearing up City Neighborhoods Hurts America, and What We Can Do About it* (New York: Random House Inc.).

Hay, R. (1998), 'Sense of place in developmental context', *Journal of Environmental Psychology* 18: 5–29.

Hummon, D.M. (1992), 'Community attachment: Local sentiment and sense of place', in I. Altman and S.M. Low (eds) *Place Attachment* (New York: Plenum Press).

Jackson, J.B. (1994), *A Sense of Place, A Sense of Time* (New Haven and London: Yale University Press).

Jiven, G. and Larkham, P.J. (2003), 'Sense of place, authenticity and character: A commentary', *Journal of Urban Design* 8(1): 67–81.

Jorgensen, B.S. and Stedman, R.C. (2001), 'Sense of place as an attitude: Lakeshore owners attitudes toward their properties', *Journal of Environmental Psychology* 21: 233–248.

—— (2006), 'A comparative analysis of predictors of sense of place dimensions: Attachment to, dependence on, and identification with lakeshore properties', *Journal of Environmental Psychology* 79: 316–327.

Kaltenborn, B.P. (1998), 'Effects of sense of place on responses to environmental impacts: A study among residents in Svalbard in the Norwegian high Arctic', *Applied Geography* 18(2): 169–189.

Kearns, R.A. and Gesler, W.M. (1998), *Putting health into place: landscape, identity, and well-being* (Syracuse, New York: Syracuse University Press).

Kianicka, S., Buchecker, M., Hunziker, M. and Muller-Boker, U. (2006), 'Locals' and tourists' sense of place: A case study of a Swiss Alpine village', *Mountain Research and Development* 26(1): 55–63.

Lewis, P. (1979), 'Defining a sense of place', in P.W. Prenshaw and J.O. McKee (eds) *Sense of Place: Mississippi* (Jackson, MI: University of Mississippi).

Low, S.M. and Altman, I. (1992), 'Place attachment: a conceptual inquiry', in I. Altman and S.M. Low (eds) *Place Attachment* (New York: Plenum Press).

Lynch, K. (1976), *Managing the Sense of a Region* (Cambridge, Massachusetts: The MIT Press).

Manzo, L. (2003), 'Beyond house and haven: Toward a revisioning of emotional relationships with places', *Journal of Environmental Psychology* 23: 47–61.

Mayes, F. (1997), *Under the Tuscan Sun: At Home in Italy* (New York: Broadway Books).

Mazumdar, S., Mazumdar, S., Docuyanan, F. and McLaughlin, C.M. (2000), 'Creating a sense of place: The Vietnamese-Americans and little Saigon', *Journal of Environmental Psychology* 20: 319–333.

Meyrowitz, J. (1985), *No Sense of Place: The Impact of Electronic Media on Social Behavior* (Oxford: Oxford University Press).

Nairn, I. (1965), *The American Landscape* (New York: Random House).

National Academy of Science. (1965), *The Science of Geography Report of the Ad Hoc Committee on Geography* (Washington: National Academy of Science – National Research Council).

Norberg-Schulz, C. (1969), 'Meaning in architecture', in C. Jencks and G. Baird (eds) *Meaning in Architecture* (New York: George Braziller).

Oritz, A. Garcia-Ramon, M.D. and Prats, M. (2004), 'Women's use of public space and sense of place in the Raval (Barcelona)', *GeoJournal* 61: 219–227.

Ouf, A.M.S. (2001), 'Authenticity and the sense of place in urban design', *Journal of Urban Design* 6(1): 73–86.

Patterson, M.E. and Williams, D.R. (2005), 'Maintaining research traditions on place: Diversity of thought and scientific progress', *Journal of Environmental Psychology* 25: 361–380.

Pred, A. (1983), 'Structuration and place: On the becoming of sense of place and structure of feeling', *Journal of Theory for Social Behaviour* 13: 45–68.

Relph, E. (1976), *Place and Placelessness* (London: Pion).

—— (1997), 'Sense of place', in S. Hanson (ed.) *10 Geographic Ideas that Changed the World* (New Brunswick, New Jersey: Rutgers University Press).

—— (2006), 'A pragmatic sense of place', [Abstract]. Senses of Place Conference, University of Tasmania, Hobart.

Shamai, S. (1991), 'Sense of place: An empirical measurement', *Geoforum* 22: 347–358.

Shamai, S. and Ilatov, Z. (2004), 'Measuring sense of place: methodological aspects', *Tijdschrift voor Economische en Sociale Geografie* 96(5): 467–476.

Shamai, S. and Kellerman, A. (1985), 'Conceptual and experimental aspects of regional awareness: An Israeli case study', *Tijdschrift voor Econ. En Soc. Geografie* 76(2): 88–99.

Stedman, R.C. (2003), 'Sense of place and forest science: Toward a program of quantitative research', *Forest Science* 49(6): 822–829.

Steele, F. (1981), *The Sense of Place* (Boston: CBI Publishing Company Inc.).

Stokowski, P.A. (2002), 'Languages of place and discourses of power: Constructing new senses of place', *Journal of Leisure Research* 34(4): 368–382.

Taylor, C.C. and Townsend, A.R. (1976), 'The local 'sense of place' as evidenced in north-east England', *Urban Studies* 13: 133–146.

Tuan, Y.F. (1980), 'Rootedness versus sense of place', *Landscape* 24: 3–8.

—— (1996), 'Space and place: Humanistic perspective', in J. Agnew, D.N. Livingstone and A. Rogers (eds) *Human Geography: An Essential Anthology* (Oxford: Blackwell Publishers).

—— (1977, Fourth Printing 2005), *Space and Place: The Perspective of Experience* (Minneapolis, Minnesota: University of Minnesota Press).

Williams, A. (2002), 'Changing geographies of care: Employing the concept of therapeutic landscapes as a framework in examining home space', *Social Science and Medicine* 55(1): 141–154.

Williams, D.R. (2007), 'Further Beyond the Commodity Metaphor: Place and Well-Being in Nature-Based Recreation and Tourism' (power point presentation at the Sense of Place and Health Workshop: McMaster University, Hamilton, Ontario, Canada).

Williams, D.R. Stewart, S.I. (1998), 'Sense of place: An elusive concept that is finding a place in ecosystem management', *Journal of Forestry* 66(5): 18–23.

Wilson, R. (1997), 'A sense of place', *Early Childhood Education* 24(3): 191–194.

World Health Organization (WHO) (1946), *Constitution of the World Health Organization* (New York: World Health Organization Interim Commission).

Chapter 3

Senses of Place and Emerging Social and Environmental Challenges

Edward Relph

In his book on the moral history of the twentieth century Jonathan Glover wrote: "Most of the time what matters is the personal and the local. But the great public disasters can strike the most unlikely places" (Glover 1999, 42). He was referring especially to genocide and war, which were the great problems of the last century, but the same principle applies equally well to the emerging social and environmental challenges of the twenty-first century. Climate change currently has centre stage, though waiting in the wings are chronic poverty, epidemics of emergent and re-emergent diseases, widespread water scarcity, ethnic conflict, the end of cheap energy, and increasing possibilities of technological error. Many of these challenges have diffuse, or global, origins and potentially widespread consequences, but there is little question that their impacts will be felt most directly in the particular places of everyday life and will have to be managed in those places.

International agreements, national policies and technical innovations should help to mitigate the effects or reduce the likelihood of public disasters associated with these challenges, but they will not entirely eliminate them, no matter how much we may wish for this. I think it is wise to assume that the various challenges, both individually and in concert, will pose serious problems at the local levels of regions, cities and neighbourhoods. By "serious problems" I mean that social and physical infrastructures could well be pushed to the breaking point, populations could be displaced by drought or flooding or intense poverty, and existing urban forms – especially low density ones based around the pervasive use of motor vehicles – could become obsolete. Such outcomes have all been identified as possible, and if they do happen it will be necessary to find ways to adapt to them and cope with them locally, in particular places, in order to maintain a reasonable quality of everyday life. And since the challenges are also global, local adaptations will have to be made with as much attention as possible to their larger contexts if they are not to exacerbate what they are trying to mitigate.

Glover (1999, 42) argues for the importance of developing individual and collective awareness of the issues around public disasters in order to limit their damage, and suggests that this is like paying taxes to fund the fire department – a sensible precautionary measure. In this chapter I pursue a similar line of thought, proposing that a need exists for what I call a "pragmatic approach to place", one that effectively lays the groundwork for finding locally appropriate ways to cope with the uncertain effects of emerging social and environmental challenges while attending always to the regional and global consequences of local actions.

The Local Impacts of Emerging Social and Environmental Challenges

It seems to be a deep human trait to expect the worst, especially around the turn of a millennium, and it is tempting to dismiss current doomsaying as just another manifestation of this tendency. Nevertheless, I think it would irresponsible not to take seriously arguments and evidence for possible of adverse social and environmental changes in the relatively near future because the consequences of ignoring them could be disastrous. The evidence begins with the history of the twentieth century. This was, in John McNeill's evocative term, a time of "screeching acceleration", a period of unprecedented growth in population, in cities, consumption of resources, and energy use (McNeill 2000, xxi–xxii). World population quadrupled to over six billion, but economic output expanded fourteen times and this enabled substantial increases in standards of living for almost everyone. However, energy consumption, carbon dioxide emissions and environmental degradation kept pace with economic expansion, and McNeill (2000, 361) concludes soberly that it appears that growth was achieved "at the risk of sacrificing ecological buffers and tomorrow's resilience". This conclusion resonates with numerous other expressions of the sense that twentieth-century growth and change may have been too much, too fast, and that its costs could well come home to roost in the present century.

In *Dark Age Ahead* Jane Jacobs (2003) wrote of the decline of family and community, of the weakness of contemporary education, of systems of taxation remote from problems, of inadequate professional self-scrutiny, and of the loss of scientific objectivity. She thought that in the absence of aggressive remedial actions the combined outcome of these, as her book title suggests, could be a swift spiraling down of modern civilization. Jared Diamond (2005), in his book *Collapse*, explores reasons, many of them environmental, for the decline of previous civilizations with the clear implication that we could be following in their footsteps. Ronald Wright has observed that many great ruins of former civilizations are monuments to "progress traps", and suggests that if we don't clean up pollution and set economic limits *now*, while we prosper, "this new century will not grow much older before we enter an age of chaos (Wright 2004, 8; 132). Martin Rees, the Astronomer Royal, examines the problems that science and technology might present in the present century, including a revived nuclear arms race, novel technologies that could get out of hand, and environmental degradation. He titled his book *Our Final Hour*, and writes that "the odds are no better than fifty-fifty that our civilization on earth will survive to the end of the present century" (Rees 2003, 8). Howard Kunstler in *The Long Emergency* (2005) offers an account of the evidence that oil and gas resources are limited, their production will soon peak, prices will then rise, and the cheap energy on which our civilization depends will come to an end with potentially disastrous consequences. The possibility of future disaster is also suggested in the Stern Review on *The Economics of Climate Change* presented to the British Government in 2006. This argues that if effective and immediate action is not taken to reduce greenhouse gas emissions the world could experience a global average temperature rise of 5C, equivalent to the change in average temperature since the last Ice Age. "Such a radical change in the physical geography of the world", it is suggested, "must lead to major changes in the human geography – where people live and how they live their lives" (Stern, Summary of Conclusions vi). The costs of this could be equivalent to twenty per cent of global GDP, tantamount to a world economic catastrophe.

This litany of warnings (and there are many others) comes from respected authorities whose arguments are well supported and whose books are widely read. Of course, and as Rees notes, none of these results are certain, but they are all possible, and this requires that serious consideration be given to the risks they pose and to precautionary measures that might mitigate them (Rees 2003, 42). In this regard it is especially important to reinforce the conclusion of the Stern Review that the various global challenges will directly affect where people live and how people live their lives. In other words, they will have a profound effect on places and the quality of life in those places.

Over the last two hundred years the quality of life improved in no small measure because the healthfulness of the places where people live was improved. For instance, sanitation and water filtration systems dramatically reduced the incidence of cholera and typhoid fever, and together with a general rise in incomes that led to improved diets and standards of cleanliness, this resulted in reduced infant mortality and longer life expectancy. It is now generally taken for granted that daily life will be reasonably clean, comfortable, secure and, for the most part, healthy.

Hurricane Katrina's impact on New Orleans demonstrated just how fragile this assumption is, and gave a foretaste of the fact that the consequences of global changes will, at least in part, be local and personal. In a similar vein, Kunstler argues that substantial increases in energy costs will undermine the comfortable security of automobile-oriented suburbs; he forecasts that "the salient fact about life in the decades ahead is that it will become increasingly and intensely local", and that "the twenty-first century will be much more about staying put than going to other places" (Kunstler 2005, 239; 263). But even if we stay put, we may be visited by epidemics of emergent or re-emergent diseases such as HIV-Aids, influenza and malaria. Sophisticated public health systems will be needed if epidemics are to be controlled, though Laurie Garrett has shown that globally public health is in a slow state of collapse because of increasing costs, growing levels of poverty and new plagues. The Interim Report of the Commission on the SARS outbreak that occurred in Ontario in 2003 repeatedly returns to the issue of the local provision of public health services and concludes that "In an age of emerging and newly emerging infectious diseases that can sweep across the world and across countries and provinces with no respect for boundaries … protection against infectious diseases is only as strong as the weakest local link" (SARS Commission 2003, 193–194). The mass international migrations of the last 40 years constitute a different type of trans-boundary mobility that have resulted in what Leonie Sandercock (2003), a social planner, has called "mongrel cities", ethnically diverse and with a latent potential for racial conflict. To reduce this possibility, she argues for therapeutic social planning that engages residents, politicians and planners in a form of place-making that aims to resolve problems on a street-by-street basis.

In summary, the challenges of the twenty-first century may be global in the sense that their causes and consequence are widely distributed, but their impacts will be felt most acutely in specific villages, towns and parts of cities – in the *places* where people live and work. Many of the challenges might be broadly considered environmental, which suggests something nebulous and remote and not worth getting too concerned about. However, it seems likely that their combined consequences will have immediate effects on people's lives. Over 30,000 deaths have been attributed to the drought in Europe in 2003 and it seems certain that future climate change and water scarcity alone will put huge stresses on the physical and mental health of individuals and communities as they try to find ways to cope. The combined and unpredictable consequences of all the challenges I have mentioned, as they interact and are superimposed in particular places, will surely exacerbate these stresses.

This is a daunting picture, one that, to borrow a compelling phrase from Mike Davis' account of the global poverty that afflicts over two billion people, can cause "the ruler's imagination … to falter" (Davis 2006, 202). Be that as it may, it would be irresponsible to do nothing to try to mitigate the possible adverse outcomes. A sensible approach for facing a complex and uncertain future is that proposed by the SARS Commission in Ontario – the adoption of a precautionary principle that recognises that actions to reduce risk do not have to wait for scientific certainty (SARS Commission 2006, 29–30). In other words, what is needed is to lay groundwork that anticipates a wide range of possibilities, yet is sufficiently flexible that it should be able to respond to very different and uncertain problems. Since many of the impacts of the challenges I have described are local, I believe that that this groundwork has to involve the development of an individual and collective awareness that balances appreciation of the complex unity of particular places with an understanding of the global character of social and environmental processes affecting them. This can be a foundation for finding appropriate local responses that can help to maintain quality of life, health and well-being in particular places in the face of global challenges. In order to advance this idea it is first necessary to be clear about what is meant by "place" and "sense of place".

Of Place and Places

Almost everything written about place notes that it is a difficult and elusive concept. It has been used to refer to a location, a social role, a building, a bit of the earth's surface or the entire earth. Aristotle defined it as a container, but for others place is a focus of meanings and experiences. In his thorough review of the history of the concept of place Edward Casey (1997) demonstrates that it was an important philosophical concept until the rise of modernism in the 17th century, when it was pushed aside in the writings of Newton, Descartes and others by the concept of space, which was considered to be more rational, universal and measurable. Thereafter place drifted in obscurity until the late 20th century. It was either assumed to be of little value for scientific explanation, or so obvious that it warranted no definition or discussion.

My own discussions of place in the 1970s, and those of Yi-fu Tuan, Ann Buttimer, David Seamon and others, were part of a revival of interest in place that was intended to redress its reduction to mere location by geographers seeking to define their discipline as a spatial science (Relph 1976; Tuan 1974, 1977; Buttimer 1980; Seamon 1979). We took the view that places are fusions of physical attributes, activities and significance, aspects of the experience of the everyday world that can be explicated phenomenologically but are inherently inaccessible to statistical analyses. David Canter, an environmental psychologist took a broadly similar view, though he approached place from the perspective of individual perceptions (Canter 1977).

For a decade or more, and with the exception of a few highlights (Eyles 1985) place slipped back into academic obscurity. However, since the early 1990s there has been a veritable explosion of writing and research that has resulted in what Williams and Patterson (2005, 361), in a comprehensive review in the *Journal of Environmental Psychology*, rather kindly call "a lack of conceptual clarity" but I am inclined to regard as serious confusion. Scholars from numerous disciplines who adopt many different theoretical perspectives have taken an interest in place, and many have attempted to adapt it to their pre-established perspectives. For example, environmental psychologists began to develop methods to investigate "place attachment", or the emotional and cultural ways that individuals connect with environmental settings (Altman and Low 1992). From an entirely different angle the economic geographer Doreen Massey proposed a large-scale view that defined place as an articulation or intersection of social relations that operate across all spatial scales. "Places," she wrote, "are not so much bounded areas as open and porous networks of social relations" with multiple, unfixed and contested identities (Massey 1994, 121). Place has since been defined as "… a concatenation of individuals connected through a set of contingent relationships" (Curry 1998, 48); as "… like space and time, a social construct" (Harvey 1996, 293); as a site of "immediate agency" (Oakes 1977), as "… the locus of desire" (Lippard 1997, 4); and as "proximal space" (Agarwal 2005, 109). Place, it seems, can be individual, social, political, emotional, ideological, aesthetic, and desirable, though perhaps not all at once.

To find a way through this confusion I turn to the work of philosophers. Edward Casey, whose work on the history of place I have already mentioned, has traced it from the origins of Western thought, to a recent revival in the writings of Heidegger, Foucault, Deleuze and Guattari, Derrida, and in Luce Irigaray's discourse on body/place. His central idea is that place is the "first of all things … we are surrounded by places. We walk over and through them. We live in places, relate to others in them, die in them" (Casey 1997, ix). Unlike many others who have written about place, Casey argues that it is fundamentally different from space and landscape. Space is an encompassing reality that allows things to be located in it or to move across it, whereas, place is "the immediate ambiance of my lived body and its history, including the whole sedimented history of cultural and social influences and personal interests that compose my life-history" (Casey 2001, 404). There is, he claims, "no place without self, and no self without place" (Casey 2001, 406). Place is situated in space, but so is everything else, and place has no privileged relationship to space. Landscape also distinguishes place from space, for there is no landscape of space, and while landscape has a horizon, space does not. Landscape is an attribute of places and can provide a context for them, but it differs from place because it is expansive – it is drawn out to a horizon – whereas place always relates to the "place-world" of the body.

Another philosopher, Jeff Malpas has gone a step further. He argues that "Place is integral to the very structure and possibility of experience"; it is a condition of existence itself, and not just a product of what is encountered in experience (Malpas 1999, 32). Like Casey, he establishes a clear distinction between place and space. While they are interconnected, place can be reduced neither to location nor to objective space, because "it is in and through place that the world presents itself" (Malpas 1999, 15). From this phenomenological perspective, "being and place are inextricably bound together in a way that does not allow one to be seen merely as an effect of the other; rather being emerges only in and through place" (Malpas 2007, 6). And, in turn, "place only appears in and through specific places" (Malpas 2007, 314). Malpas sees human lives and activities as enmeshed with the world in and through places in a dynamic relationship that is constituted through changing juxtapositions, displacement, activity and movement. A place, he argues (Malpas 1999, 160; 172; 192), is a complex unity, an open, boundless region that can turn outwards to reveal other places or inward to reveal its own character.

In effect, Casey and Malpas turn most definitions of place on their heads. Except in some very trivial senses, it is not a bit of space, nor another word for landscape or environment, it is not a figment of individual experience, nor a social construct, and it is certainly not susceptible to quantitative excavation. It is, instead, the foundation of being both human and non-human; experience, actions and life itself begin and end in place.

Sense of Place and Sense of A Place

The distinction yet connection that Malpas makes between "place" and "places" is subtle and important. Place is an abstraction from the specificity of the places that we encounter, but even in its abstractness it always implies specificity. Conversely, it is only through the specificity of places that the fundamentally ontological character of place is manifest. This suggests that sense of place has three aspects, one ontological, one focused on a particular place, and one that opens out to acknowledge differences and interactions between many places. The first has to do with how we grasp being and its relationship to the world; the second provides roots and security; the third mitigates against tendencies to parochialism and exclusion and puts local matters into a larger context.

Differentiating "Sense of place" from "sense of ***a*** place", because the first term refers to the critical ontological awareness that existence is always placed and unavoidably engaged with the unities and differences of the world, while the latter term refers to the faculty by which we identify the properties of specific places and appreciate the differences. This faculty is synthetic, or more strictly synthaesthetic, because it combines seeing, hearing, smelling, and touching with memory, responsibility, emotions, anticipation and reflection. In the first instance, following from Casey's idea that place begins in the ambiance of one's body, both sense of place and sense of a place are individual, but they are also intersubjective, so an individual's experiences of a place are always framed by and contribute to a social context and shared language. This means that, for example an individual's sense of a place is reinforced by belonging to a community in that place, something that is often reflected in shared pride and responsibility for place.

Sense of a place is enormously varied and is not constant over time. Some individuals are deeply rooted and gain a strong sense of well-being and security from knowing and being known where they live and work; others find this sort of close familiarity oppressive and can imagine nothing better than to escape from it, to travel to many different places. For those mostly concerned with making money, or conversely for the deeply poor finding a way to survive, sense of a place is unlikely to be at the forefront; on the other hand there are many cases of individuals and groups who are willing to die to protect their places, it is their primary need in spite of economic or other hardships. More generally, the relationship between people and the places they live must have varied over the course of history because of the huge changes in mobility. For example, in the 1880s my grandparents moved from different parts of Britain to a small village in South Wales and spent the rest of their lives there, rarely traveling away. My grandfather ran a small building firm, and built the house he and my grandmother lived in until they died. Their sense of that place must have been deep and intensely focused. I grew up in the same village, spent several years in the house my grandfather had built, then moved away to university, and I have subsequently lived in several cities on different continents, travel frequently to other countries to attend conferences or for vacations, and use the internet almost every day to tap into the global flow of information and to stay in touch with relatives and colleagues who live thousands of kilometers away. My place experiences, unlike my grandfather's but not unlike those of hundreds of millions of others, are now spread-eagled across the world.

It would be easy to assume from this that my sense of particular places is somehow less rooted and therefore deficient in comparison with that of my grandfather and his contemporaries. Realistically there is no way of knowing this and in any case there is no way of going back, of undoing modern means of communication and returning to a deeply rooted, local way of life. Nevertheless, it is important to understand how such changes might inform the ontological sense of place. While my grandfather's range of experience was rooted, it was also relatively narrow. Rootedness is generally considered to be positive, something that contributes to well-being and quality of life because knowing and being known somewhere provide security and dependability. However, concomitants of narrow place experience are parochialism, exclusion, and a tendency to reject unfamiliar differences. By comparison with those of my grandfather, my experience and sense of places are relatively shallow and fleeting – a few days here, a few years there. They are, however, widely distributed, and encompass a wide variety of places and cultures. The geographer Paul Adams (2005) uses the term "extensibility" to describe how everyday lives are now constantly connected to distant places. If I may borrow his term, contemporary sense of places is "extended". It is characterized by breadth rather than depth of experience, a breadth that offers opportunities for comparisons and an appreciation of place differences that would have been possible for my grandfather only as an act of imagination.

Doreen Massey would have regarded my grandfather's place experience as bounded and nostalgic. Her preference is for a sense of place that grasps places as nodes in flows of social and economic relations. However, when she describes "a real place" – Kilburn in London, where she was living at the time she wrote – she recognizes the need to connect her place attachment to Kilburn with the time-space compression of the present age (Massey 1994, 151–152). This she achieves by proposing what she calls a "global" or "progressive" sense of place that looks through local details to grasp their connection to global patterns and processes.

A progressive sense of place is, in my interpretation, a version of what I called above "sense of places". While I have reservations about Massey's spatially abstracted idea of place, this is nevertheless a powerful idea, one that connects with extensibility, acknowledges the realities of contemporary geographical experiences and yet grasps the importance of place attachment. What it does not do is to connect with an ontological "sense of place". To make that link it is necessary to consider Jeff Malpas' argument that place is integral to existence and always opening out to reveal other places or turning inward to reveal its own identity. From this perspective, sense of place has to do with our grasp of being in the world, while sense of *a* place is about attachment to a particular place, and sense of *places* has to do with the appreciation of relationships and differences between many places.

This combination of sense of places and ontological sense of place is not some sort of fine art extra to be entertained after economic and practical matters of health, economics and society have been considered. On the contrary, I believe it is an existential foundation for individual and communal well-being. It is also the basis for a pragmatic approach to place that can inform the local adaptations needed to deal with environmental and social challenges of contemporary society.

Well-Being and a Pragmatic Approach to Place

I have argued that there is strong evidence that serious social and environmental challenges are emerging at the beginning of the twenty-first century. While many of these are global in the sense that they have dispersed causes and will affect much of the planet, their impacts will be experienced most strongly at a local level, in diverse places where they have the potential to disrupt everyday life and seriously undermine the well-being of individuals and communities. Hippocrates suggested that there is connection between place and healthfulness, and it is still widely accepted that strong connection between person and place contributes positively to well-being, and therefore to the health of individuals and communities (Williams 1999). This idea is strongly reinforced by the argument that places are the foundations not just of well-being but of being itself. The depth and diversity of this relationship are disclosed through sense of place. It is in places that the impacts of the various challenges will be most immediately experienced, and it is through sense of place that the practical possibilities for dealing with these impacts will have to be developed. Moreover, because of the diversity and uncertainty of the challenges, standardized approaches for mitigation and adaptation are unlikely to be effective; different places will require different strategies.

Not for everyone perhaps, but for many people, sense of a particular place provides a feeling of belonging and of security, of being part of a community that shares responsibilities and provides mutual support through informal social networks. This applies most clearly at a very small scale, such as part of a block or a street, where residents pay attention to the needs of their neighbours, perhaps taking care of them when they are ill, or helping with child-minding and with chores. The benefits of these sorts of social relationships lie both in being a responsible citizen, and in receiving the assistance of others when it is needed. They are such normal components of relationship to place and community that they often go unremarked, so they are perhaps most apparent in their absence. This is certainly suggested by Marc Fried (2000) when he observes that cases of forced relocation or displacement, which disrupt systems of mutual support, are considered by to be among the most severe of all psycho-social impacts.

Sense of a place also involves familiarity with the physical environment of somewhere – knowing its weather, its seasons, the plants, hills, buildings and streets, and being attentive to changes in them. This type of observation and watching might take the form of *flaneurie* – a sort of idle reflection about what is going on, but more normally it involves a measure of responsibility, assessing changes and how they might affect the quality of life in this place, and then possibly acting to facilitate or impede them. At the scale of an urban neighbourhood or small town these changes can be assessed in terms of what Kevin Lynch in *A Theory of Good City Form* (1981) described as "performance criteria." These criteria include *vitality* – the degree to which somewhere promotes or diminishes life (through good water quality and air quality and so on); *imageability* – the degree to which an environment is coherent and legible; *control* – the degree to which citizens have control over the environments in which they live and work; *efficiency* – how well it works; and *justice*– how equitably the benefits and costs are distributed.

There has always been a practical aspect to the sense of a place that has involved the sorts of criteria Lynch describes. In other words, this is not just a passive and receptive sense – to live somewhere requires, at the very least, maintaining and generally taking some practical responsibility for the way it is. And it also requires some degree of designing, building and making transformations to it. For some individuals this may not extend much beyond their own property and the occasional renovation. For others some combination of skills, interests, social commitment and professional orientation can extend responsibility to the neighbourhood or city where they live, and, in effect, they take on the role of place-maker.

These practical and responsible aspects are the means by which the sense of a place is translated into the built forms of that place, and they are an essential component of creating distinctive. But if they become too narrowly proscribed they can contribute to exclusionary attitudes, and it is precisely because of this that an *extended sense of places* is so important; it leads to an appreciation of differences and challenges insularity. An extended sense of places is also to bring environmental and other challenges in perspective. David Uzzell (2000), in a study of attitudes to global environmental problems, has noted a spatial disconnection between perception of the severity of problems and responsibility for them. Environmental issues are perceived to be most serious globally, at a scale where people feel most powerless to do anything about them; at the local neighbourhood level, for which they feel the greatest responsibility, environmental problems are perceived to be minor. This suggests that, while the psychological foundation for local responsibility already exists, it is necessary to promote an extended sense of place in order to establish an understanding of the strong links between local actions and global consequences.

The practical, local responsibility implicit in sense of *a* place, combined with the insights of an extended sense of places, constitute the foundation for what I call "a pragmatic approach to place." Pragmatism was originally proposed in the late nineteenth century to resolve endless abstract debates about theoretical issues in the philosophy of science. William James (1973(1906)) defined it neatly as a way of looking away from theoretical principles towards consequences and facts. In the last forty years this idea has been expanded by philosophers such as Richard Rorty (1982; 1999) and Stephen Toulmin (2001) as a basis for coping with social and political indecision about issues that are not susceptible to rational solutions, including many matters of justice, civil rights, inequality and gender. Neo-pragmatic philosophers advocate getting back in touch with consequences of decisions, returning to issues of everyday life, and choosing reasonable strategies on the basis of the best available information. Toulmin (2001, 213) cites *Médécins sans Frontière*s as an example of a group that uses available expertise and medicines to do what it can to deal with health problems of individuals and communities in uncertain conditions in the developing world. There are, of course, complex economic and political issues behind these problems that are subject to many different interpretations, and these cannot be ignored, but the immediate aim of the doctors involved is to do what they can at a local and individual scale to prevent things getting worse and to improve people's health. This is a sage, precautionary and pragmatic model that can be used for coping with climate change, water scarcity, the end of cheap energy and the various other challenges I have mentioned.

A pragmatic approach to place is precautionary in two ways. First, local places can contribute incrementally to the solution of large-scale problems, most obviously by not exacerbating them. It is, however, important to realize that even the most aggressive and well-intentioned local actions cannot prevent, for example, the increasingly extreme weather associated with climate change, or the escalation of energy costs, because the processes that cause these have already been set in motion. In other words, it is too late for prevention, so it is the possibilities for local mitigation that should command our attention. Secondly, different local places, that is to say municipalities and the neighbourhoods within them, will inevitably be on the receiving end of environmental and social problems, will experience them differently, and will have to find locally appropriate ways to cope with them. In other words, a pragmatic approach to place is precautionary in that it establishes a basis for appropriate and necessary actions by balancing local attachment against an appreciation of distant causes and consequences.

A key term in this is "appropriate and necessary actions". By this I mean adaptations that fit well with the specific values and attributes of a place because they emerge from individual and community attachment to a place, without losing sight of broader regional and global contexts. Such actions simultaneously have to open out to the world, in other words to be progressive and extended, and to look inward to the social, physical and historical sources of local identity. This is a difficult balance to achieve, but it is necessary in order to frustrate the opposing tendencies either to adopt *ad hoc* local and reactionary measures that fail to acknowledge their contribution to global issues, or to seek universal solutions that will suppress local identity and responsibility. It is helpful to understand this balance in terms of Lynch's performance criteria that I mentioned earlier. A pragmatic approach will, in the face of adverse changes, aim to maintain the vitality of a place, to protect qualities that contribute directly to the health and well-being of its inhabitants, to protect distinctiveness and to use it as a basis for developing local measures for mitigation and adaptation, and to do this in the most efficient and equitable way possible. It will do this by recognizing that those who live and work in a place have a large measure of responsibility for how it responds to challenges. National and municipal governments may provide overarching guidelines and establish policies, but if the challenges are as serious and diverse as I have suggested, much of the responsibility for adapting to them will fall to local, place-based communities.

None of this is entirely new. In some respects it represents a self-conscious return to the sort of situation that seems to have existed out of necessity in pre-modern places, before the technological changes of the nineteenth and twentieth centuries, when difficulties of transportation ensured that local materials were used for construction, and made it necessary for local communities to feed and otherwise fend for themselves. The difference now is that both collective awareness about the local consequences of social and environmental challenges, and the place-based responsibility for coping with these, have to be deliberately cultivated.

To be effective, a pragmatic approach to place has to infiltrate political and planning thought and practice at the local, municipal level, because this is scale at which place-attachment is most apparent and where citizens and cities have some chance of affecting their own destiny and their own quality of life. However, the consequences of climate change, the gap between rich and poor, the problems of mongrel cities and the other challenges confronting the twenty-first century, cannot be addressed only by actions at the local level. These are problems that have an international character, and they will require multilateral agreements and national policies to direct strategies for adaptation and mitigation. But even for these a pragmatic approach to place has value, because it will help to ensure that these policies are capable of responding to diverse local circumstances. In dealing with places a single, standardized policy will never fit all.

Conclusion

The approach to place that I have proposed is essentially a humanistic one that balances individual and local concerns with a broad critical understanding of shared values and interests. It will neither be easily achieved nor easily maintained. Edward Said has observed that "the disheartening part" about a humanistic view is that the more its importance for understanding differences has become apparent through critical studies of culture, the less influence it seems to have (Said 2003, xxii–xxiii). Humanism, he suggests, is first of all the attempt to dissolve Blake's "mind forg'd manacles" so as to be able to use one's mind for reflective understanding and disclosure. From a phenomenological and humanistic perspective, place is inseparable from being, but this view has long been suppressed either by the delusion that measurable space takes precedence over place, or by rationalistic approaches to planning that have treated place as something that is marginal to goals of profit and efficiency.

The importance of a humanistic, pragmatic understanding of place is now slowly being rediscovered, both as an aspect of the collective awareness that everywhere is part of the whole, and as a precautionary foundation for finding ways to cope with adverse change. This rediscovery revolves around three related senses of place – a sense of the inextricable ontological connection between being and place, a sense of the unique qualities of a particular place, and an extended sense of the connections between many different places. Whether this is actually referred to as "a pragmatic approach to place" is not important. What matters is the recognition that ways of dealing with global issues have to be balanced against senses of place, and vice versa, in order to find reasonable, locally responsible strategies that will allow places to weather the worst of the challenges that could well confront them in the near future. And in the delightful event that I am utterly mistaken about the local impacts of these social and environmental challenges, it seems to me that no great harm will have been done by actively promoting this approach to place. On the contrary, it can only hasten the rediscovery that place is integral both to being and to well-being.

References

Adams, P. (2005), *The Boundless Self: Communication in Physical and Virtual Spaces* (Syracuse: University of Syracuse).

Agarwal, P. (2005), 'Operationalising Sense of Place as a Cognitive Operator for Semantics in Place-Based Ontologies', in A.G. Cohn and D.M. Mark (eds), *Spatial Information Theory* (Berlin: Springer-Verlag).

Altman, I. and Low, S. (eds) (1992), *Place Attachment* (New York: Plenum Press).

Buttimer, A. (1980), *The Human Experience of Space and Place* (London: Croom Helm).

Canter, D. (1977), *The Psychology of Place* (London: The Architectural Press).

Casey, E. (1997), *The Fate of Place: A Philosophical History* (Berkeley: University of California Press).

—— (2001), 'Body, Self and Landscape: A geophilosophical enquiry into the Place-World', in P. Adams, S. Hoelscher, K.E. Till (eds) *Textures of Place: Exploring Humanist Geographies* (Minneapolis: University of Minnesota).

Curry, M. (1998), *Digital Places: Living with Geographic Information Technologies* (London: Routledge).

Davis, M. (2006), *Planet of Slums* (London: Verso).

Diamond, J. (2005), *Collapse* (New York: Viking Books).

Eyles, J. (1985), *Senses of Place* (Warrington: Silverbrook Press).

Fried, M. (2000), 'Continuities and Discontinuities of Place', *Journal of Environmental Psychology* 20: 193–205.

Garrett, L. (2000), *The Betrayal of Trust: The Collapse of Global Public Health* (New York: Hyperion).

Glover, J. (1999), *Humanity: A Moral History of the Twentieth Century* (Pimlico: London House).

Harvey, D. (1996), *Justice, Nature and the Geography of Difference* (Oxford: Blackwell).

IPCC (2007), *Fourth Assessment Report Climate Change 2007, Climate Change Impact, Adaptation and Vulnerability* Summary for Policymakers.

Jacobs, J. (2003), *Dark Age Ahead* (New York: Random House).

James, W. (1967) (1906) 'What pragmatism means', in *Pragmatism and Other Essays* (New York: Washington Square Press).

Kunstler, J.H. (2005), *The Long Emergency: Surviving the Converging Catastrophes of 21st Century* (New York: Atlantic Monthly Press).

Lippard, L. (1997), *The Lure of the Local: Sense of Place in a Multi-Centred Society* (New York: The New Press).

Lynch, K. (1981), *A Theory of Good City Form* (Cambridge: MIT Press).

Malpas, J. (1999), *Place and Experience: a Philosophical Topography* (Cambridge: Cambridge University Press).

Malpas, J. (2006), *Heidegger's Topology* (Cambridge: MIT Press).

Massey, D. (1994), *Space, Place and Gender* (Cambridge: Polity Press).

McNeill, J. (2000), *Something New Under the Sun: An Environmental History of the Twentieth-Century World* (New York: W.W. Norton).

Patterson, M. and Williams, D. (2005), 'Maintaining Research Traditions on Place: Diversity of Thought and Scientific Progress', *Journal of Environmental Psychology* 25(4): 361–80.

Rees, M. (2003), *Our Final Hour: A Scientist's Warning How Terror, Error and Environmental Disaster Threaten Humankind's Future* (New York: Basic Books).

Relph, E.C. (1976), *Place and Placelessness* (London: Pion).

Rorty, R. (1982), *Consequences of Pragmatism* (Minneapolis: University of Minnesota Press).

—— (1999), *Philosophy and Social Hope* (London: Penguin).

Said, E. (2003) (1979), *Orientalism* (New York: Vintage Books).

SARS Commission (2004), *Interim Report: SARS and Public Health in Ontario* (The Cambell Commission), Province of Ontario.

SARS Commission (2007), *Spring of Fear* (The Campbell Commission) Province of Ontario.

Sandercock, L. (2003), *Cosmopolis II: Mongrel Cities of the 21st Century* (London: Continuum).

Seamon, D. (1979), *A Geography of the Life-World: Movement, Rest and Encounter* (London: St. Martin's Press).

Stern, N. (2006), *The Economics of Climate Change (The Stern Review)* (London: HM Treasury).

Toulmin, S. (2001), *Return to Reason* (Cambridge: Harvard University Press).

Tuan, Y.-F. (1974), *Topophilia: A Study of Environmental Perception, Attitudes and Values* (New Jersey: Prentice Hall).

Uzzell, D. (2000), 'The Psycho-Spatial Dimension of Global Environmental Problems', *Journal of Environmental Psychology* 20: 307–318.

Williams, A. (ed.) (1999), *Therapeutic Landscapes: The Dynamic Between Health and Place* (Lanham: University Press of America).

Wright, R. (2004), *A Short History of Progress* (Toronto: House of Anansi).

Chapter 4

Holistic Paradigms of Health and Place: How Beneficial are they to Environmental Policy and Practice?

Ingrid Leman Stefanovic

In one of his story collections, author Wendell Berry describes the case of a farmer who decides to impersonate medical personnel, in order to covertly remove his dying father from the sterile efficiency of an intensive care unit so that the old man can die at home on his farm, on his land (Berry 1992). The story is moving at many levels but it certainly also speaks to our theme of health and place. As the machines, tubes and needles prolong his life in the hospital environment, it becomes clear that dying with dignity can only happen for the father within a sense of place and belonging that is part of being-at-home.

Certainly, the story reveals intertwining truths that a sense of place consists of more than mere geographical location and that health and well-being consist of more than mere physiological survival. In the words of some phenomenologists, health and place are best described as holistic, ontological structures of being human and, within such a description, they exceed reductionist parameters. This chapter begins by presenting some of these reflections, first on the phenomenon of place and then, on the issue of health, in order to suggest that there is something special *in kind* when it comes to capturing the richness of these concepts.

Beyond this description, I then discuss the challenges that present themselves when we seek to measure indicators of health and sense of place. I suggest that qualitative research methods must supplement traditional quantitative measures and I finally present some findings from a case study that may be useful as an example of how qualitative studies can aid in determining policy.

The Phenomenon of Implacement

There have been a number of major texts published in recent years that build on phenomenologist Martin Heidegger's reflections on dwelling as a fundamental structure of being human and extend these reflections to the notion of place (Malpas 1999; Norberg-Schulz 1985; Relph 1976; Seamon and Mugerauer 1989; Stefanovic 2000). Edward Casey's *Getting Back into Place* is one such text that does a particularly thoughtful job of showing how "to be is to be in place" (1993, 14). Inasmuch as we exist, we exist *somewhere* – both temporally and spatially – and, as Heidegger himself has shown, that fundamental implacement is the very condition of the possibility of our comporting ourselves within the world (Heidegger 1971). Building on Heidegger's writings, Casey explains that "place serves as the *condition* of all existing things." This means that, far from being merely locatory or situational, place belongs to the very concept of existence" (1993, 15).

To put it another way, we do not first exist and then, secondarily, exist *somewhere*. On the contrary, existence requires spatial and temporal placement. If ontology is a description of our manner of being-in-the-world, then we conclude with Heidegger that "to be in place" is part of the fundamental *ontological* structure of being human (Heidegger 1962). Certainly, such ontological implacement is often taken for granted: it is not some "Thing" that we explicitly and logically navigate on a daily basis. On the contrary, our spatio-temporal horizons shape our understanding of things within the world and, to that extent, implacement "grounds" our everyday activities – not as some "deeper," underlying Meta-physical Foundation of existence but, on the contrary, as an *Abgrund* – a "ground without grounds" that, being other than any single, circumscribable entity, serves as the very condition of the possibility of the appearance of discrete elements that we encounter and negotiate in our practical dealings with the world.

While there has been a growing body of literature in this area of thought, there have certainly also been some skeptics. Environment and Behaviour (EB) researcher, Amos Rapoport, has argued that sense of place is a vague term that is never clearly defined. In his view, when attempts are made to delineate the concept, those attempts are either illogical or so general as to be meaningless (1994). As far back as 1983, some critics argued that the dialogue around place was simply a passing fad (Clay 1983, 110).

While the phenomenon of place has withstood the test of time, the frustration around locating a universal definition is not trivial. Nevertheless, I argue that it is precisely the difficulty of arriving at a clearly delineated definition that point to the very richness of the concept of place. Let me explain.

In one of his more readable essays, Martin Heidegger distinguishes between what he calls "calculative" and "meditative" thinking (1966). Heidegger famously argues that the contemporary age of science and technology has been "in flight from thinking" (Heidegger 1966, 45). Certainly, he acknowledges that a certain approach to thinking – the calculative mode that "computes ever new, ever more promising and at the same time more economical possibilities" – has defined many scientific accomplishments and, to this extent, it "remains indispensable" (Heidegger 1966, 46). Calculative thinking remains calculation, according to Heidegger, even if it does not work with numbers explicitly. Its defining characteristic is that it relies upon clearly delineated, empirically measurable *things*.

The father of modern philosophy, Rene Descartes, defined the human being, in his "First Meditation," as a "*thing* that thinks" (1968, 106–107). This description is no small matter: consider how many people today consider the human body to be equivalent to a machine, whose component parts must be understood in order to treat disease. Both the reductionism of such an interpretation of the human body as well as the interpretation of embodiment in purely physical, empirically accessible terms, can be traced back to Aristotle, who first distinguished in his *Metaphysics* Delta Book 7, between Real Being (*ens reale)* and Conceptual Being (*ens rationis*.) (See

Barrett 1958, 288ff). According to Aristotle, "Real Being" referred to that which was *actual*, that is, to that which possessed concrete existence and whose parameters were subject to empirical measure. "Conceptual Being," on the other hand, pointed to the existence of entities but those which were only "in the mind."

So, for example, the City of Toronto as *a particular place* possesses, on Aristotle's reading, "Real Being" to the extent that we can (and do) geographically delineate its physical borders. When we talk about a "sense of place" (or lack of it!) within the City of Toronto, we are, in Aristotelian terms, referring to something that exists, but only as a "concept."

Immediately, however, we begin to recognize the limits of the Aristotelian distinction. Is implacement no more than a geographical reference? Is "sense of place" merely an empty concept (and one which, in Rapoport's terms, is vague and illogical?) Does sense of place not, more correctly, refer to a *lived experience* that is more meaningful than that which is found in a conceptual abstraction or mere propositional truth?

Existential phenomenologist, William Barrett, traces the Aristotelian understanding of Being through to a Mediaeval tradition that enlarged the interpretation of "Conceptual Being" beyond concepts and propositions, to include also privations, absences and entities that cannot be subjected to empirical measures (Barrett 1958; Stefanovic 2000). St. Thomas Aquinas offered up the example of blindness: in the true spirit of calculative thinking, Aquinas argued that the eye and the cataract were measurable entities and, to that extent, were "real." The experience of "not-seeing" existed, on the other hand, only to the extent that "the eye does not see" was a true proposition. The experience of blindness, in the Thomistic paradigm, was reduced to mere conceptual being (Barrett 1958, 288–289; Stefanovic 2000, 30).

The limits of such an interpretation of the experience of blindness begin to be clear: if I were to suddenly be blind, is this merely a conceptual reality for me? Certainly, this is no more true than the attempt to reduce the City of Toronto's sense of place to something that is "purely in the mind." Something is missing here.

In fact, when we seek to understand reality purely in terms of things and quantifiable objects, on the one hand, relegating non-objects or human experiences to ephemeral conceptual states, on the other, we risk losing something essential about the meaning of life. Subjecting all realities to calculative indicators means that taken for granted contexts and non-quantifiable values, relationships and lived experiences are ultimately less important than that which can be configured within empirical measures.

Heidegger believed that, beyond calculative thinking, we also had to recognize the validity of what he called "meditative" thinking – but which I prefer to call "originative" thinking, since this term indicates both the significance of origins, as well as the fundamentally creative element of human understanding of the world (Heidegger 1966; Stefanovic 2000). Heidegger reminds us that, in addition to the calculation which underlies so much of today's scientific and technological accomplishment, we must not forget to "contemplate the meaning which reigns in everything that is" – a meaning that is never fully revealed but which we can begin to contemplate if we remain open to the mystery that "hides itself just in approaching us" (Heidegger 1966, 46, 55).

In this vein, when Rapoport criticizes place theorists for being too "vague," he is perhaps right to the extent that the term exceeds specific, calculative parameters. At the same time, he is wrong, if he judges such "vague" reflections to be meaningless. When we speak of a "sense of place," we recognize that there is a depth to the phrase which cannot be captured either in reductionist, physical measures, or in ephemeral "concepts." There is an "ever-more" to the notion of place that emerges between the interplay of spatio-temporal location, on the one hand, and human experience, on the other. Metaphysically collapsing the richness of human experience to the dualism of either objective, measurable entities or conceptual states "in the mind" closes the door to an understanding of the relationship *between* empirical realities and human understanding. It is this *ontological* dimension of place that captures the human-environment relation and that deserves our attention. As we shall see in the next section, it is similarly an ontological interpretation of health that serves to confound simplistic definitions of well-being.

Embodiment and Health as Ontologically Embedded

As we reflect on place, we must also come to terms with the essential belonging of implacement to the phenomenon of embodiment. It is only as embodied creatures that we dwell. Place is experienced bodily. In fact, in the words of phenomenologist Bernd Jager, "to enter and come to inhabit a place fully means to redraw the limits of our bodily existence to include that place – to come to incorporate it and to live it henceforth as ground of revelation rather than as panorama. An environment seen thus is transformed into a place which opens a perspective to the world" (Jager 1985, 220).

Jager is suggesting that to be embodied means, at the same time, to be implaced. In that regard, place is more than a separate, distant "panorama" but defines the very meaning of ourselves as embodied. He picks up on the classic writings of phenomenologist Maurice Merleau-Ponty, who described the mystery of the body as belonging to the visible, inasmuch as it can be seen – but it is also, in some sense, hidden in constituting the very source of vision itself (Merleau-Ponty 1968). When my hand grasps an item, I rarely view the hand as an object, in and of itself. Rather, it recedes to become the taken-for-granted means whereby I access the book or the cup upon which my attention is focused. Similarly, inhabited environments, wherein I truly dwell, are not simply explicit objects but are the prethematically grasped extensions of myself: my home is precisely the place where I can be myself, immediately orienting myself in a way that I cannot in a strange hotel room because home becomes a pre-thematic extension of my embodiment.[1]

According to Jager, "to approach inhabitation in this way means to be able no longer to make such a radical distinction between flesh and matter, between bodies and mere things. Bodily existence floods over into things, appropriates them, infuses them with the breath of life, draws them into the sphere of daily objects and concerns" (Jager 1985, 219). In this process, the body becomes transparent, no

1 To understand something "prethematically" means to grasp it immediately, "pre-consciously," prior to explicit, thematic representation or description.

longer functioning as a "confronted" thing but, rather, it recedes to reveal a visible world (Jager 1985, 218–219).

It is this "transparency" of embodiment that intrigues me. In many ways, the body is not an object of reflection but, as embodied, I am essentially something other than a thing among things. In fact, the body becomes *no-Thing* in enabling place to emerge just as sense of place is more than any *one thing* or even a mere accumulation of *things*. Moving beyond the calculative paradigm, we recognize that an ontological description of place, embodiment and (as we shall see) health requires a sensibility to non-reductionist and non-empirical measures because these phenomena reveal themselves no longer as confronted objects, capable of being quantified.

The transparency of the body in our everyday dealings and its implications in reflecting on the phenomenon of health is best described by hermeneutic philosopher, Hans-Georg Gadamer. In his book, *the Enigma of Health: The Art of Healing in a Scientific Age*, Gadamer points to the mystery of embodiment when he recognizes that it is precisely when health is *present* that it is *absent* from explicit view. "Health is not something that is revealed through investigation but, rather, something that manifests itself precisely by virtue of escaping our attention...It is not something which invites or demands permanent attention. Rather, it belongs to that miraculous capacity we have to forget ourselves" (Gadamer 1996, 96). Good health, according to Gadamer, is the primary, taken-for-granted ontological condition of "being ready for and open to everything" (Gadamer 1996, 73).

In many ways, good health means that the body is forgotten as an object, becoming instead the condition of the possibility of encountering a world. Perhaps this is why it is easier to confront specific symptoms of ill health, than it is to define a holistic sense of well-being. Similar findings apply to the broader notion of environmental health. Clearly, human welfare is often directly or indirectly linked to a healthy planet. David Rapport, former chair of Ecosystem Health at the University of Guelph, Ontario, Canada, acknowledges, however, that we have greater difficulty in coming to a universal definition of what constitutes a healthy and sustainable ecosystem, than to pointing to problems of ecosystem dysfunction (Rapport 1997, 45). Good health is a holistic notion, much as sense of place. Trying to capture such notions within reductionist, calculative paradigms misses the essence of the lived experience of both human, as well as ecosystem health.

Certainly, non-phenomenologists have also recognized the benefits of understanding the phenomenon of health from a holistic perspective, one that moves beyond narrow positivist enquiry. Baird Callicott, for example, acknowledges that while "ecosystems are not as clearly bounded as are most organisms," there is merit to developing a "synthesizing" understanding of ecosystem health (1997, 134–135).

Thomas Burke similarly argues that "Balkanized" approaches to environmental decision making, limit the ability of national agencies to respond to environmental challenges (Burke 1996, 95). He suggests that a "public health" perspective moves beyond single-hazard, pollutant-specific studies that have traditionally employed toxicological models and top-down regulatory standards as sole strategies for environmental decision making. A broader, more holistic "public health" perspective will, on Burke's reading, allow for more diverse tools, that are population and community-based in nature, and that will offer a "more 'bottom up' decision-making process" (Burke 1996, 99).

Perhaps it is not so much a question of choosing between approaches as supplementing the traditional calculative approaches to issues of health with more holistic perspectives. In Burke's words, "by adopting a combined approach, environmental decision making might move beyond the pollutant and media-specific approach toward a population-based approach to community health" (Burke 1996, 100).

If Burke is right and we are wise to consider both traditional reductionist methodologies as well as more "holistic" approaches to environmental health, the question remains: what tools are available as indicators of environmental health, under such a rubric? A related problem is how one can move from non-quantitative approaches to policy recommendations, when traditional quantitative measures are lacking. I consider these problems in the following section and turn to a case study, in the hope that it might prove useful in these reflections.

Qualitative Methodologies in the Study of Place and Health

Still immersed within a calculative paradigm, our society frequently belies a continuing underlying belief that quantitative measures are superior to the qualitative. Even when trying to elicit information about something as elusive as peoples' values and preferences, the tools utilized by researchers frequently aim for quantitative indicators. In fact, "perhaps the most well-known of the formal approaches [to preference construction and elicitation] is termed multiattribute trade-off analysis, which involves an interview between an analyst and a decision participant. The result," as the National Research Council proudly reports, "is a mathematical statement comprising a utility or value function that could be used to evaluate every possible alternative within a range of consequences used in the interview process" (NRC 2005, 33).

While there may be some significant merit in such approaches, my own sense is that qualitative methodologies are far better positioned to reflect complex aspects of human behaviour and perceptions. There are a number of reasons for this conclusion. First, human beings, in principle, are themselves more than delimited objects that can be empirically measured in the same way as other entities. Second, even quantitative surveys that require human feedback often predetermine categories of responses that oversimplify the situation, merely in order to elicit a mechanism for counting. Finally, qualitative research methodologies have come a long distance over the last decades, evolving to be more sophisticated and sensitive to complexities of human understanding, at the same time also developing new standards of rigor in the research. While not specifically quantitative in nature, these qualitative studies (particularly in the field of health) have provided new information, otherwise unavailable, as new standards of research integrity have also been developed (See, for example, Crabtree and Miller 1999; Creswell 1998).

By way of illustration, let us consider some of the findings that have emerged from a study under my direction, aimed to evaluate values and perceptions of the

Figure 4.1 Waterfront Trail

Source: http://www.waterfronttrail.org/

Lake Ontario Waterfront Trail.[2] The trail itself is an innovative experiment, originally launched through the leadership of the former Mayor of Toronto, the Hon. David Crombie, together with the Waterfront Regeneration Trust. Running from Niagara to Trenton, through urban, rural, conservation areas, parkland and small towns, the trail allows for hikers and bikers to travel over 750 km along designated trail areas, often adjacent to the lake itself. (See Figure 4.1, found at: http://www.waterfronttrail.org/)

Our own project was aimed to elicit information from those who had traveled the trail, about their perceptions of its success, its challenges and its value overall as an environmental experience. The work relied upon qualitative research methods and included interviews of 25 "end-to-enders," that is, those who had either hiked or biked the travel from one of the trail to the other; as well as work with groups of school children, aged 9–14, in areas principally covering the western municipalities of the Greater Toronto Area.

While "health" explicitly was not a predominant theme amongst the adults interviewed for this study, the concept certainly did emerge during our conversations with them. The strategy employed in the semi-structured interview process was

2 This study was originally funded by the Social Sciences and Humanities Research Council (SSHRC) but additional funding was provided over the last few years by the Faculty of Arts and Science, University of Toronto.

meant to elicit spontaneous descriptions of the trail, with minimal disruption of the conversations by the interviewers. What was significant about this part of the research in terms of the purposes of this chapter, was the fact that, when it did emerge as a topic of discussion, *health itself was intimately connected with the theme of place.*

Interviewees were asked to provide five-word descriptors of the trail and "healthy" was one of them. What were some of the elements that contributed towards this vision of the trail as a healthy environment? One respondent felt that the trail is "colourful. It's varied. It's good for you athletically. Healthy. It's sort of uplifting. And I think it's visually varied." A number of participants described the trail as wonderfully "diverse" as it traversed urban, rural, park and small town environments.

The Lake Ontario Waterfront Trail emerged as "healthy," as well as a "place of variety" and "freedom," "friendly," "exciting," "peaceful," "safe," "uplifting," "beautiful," "historically revealing" and "connected with the lake." The most frequently cited "positive criterion" was the fact that the trail was "new and unfamiliar," reflecting a "sense of contrast/juxtaposition," where visitors could interact with the locals, purchase local products, experience local events and engage in a "special nature experience" at the same time. The trail as a "healthy" place to be was a place of "shared history and place," "accessibility," "diversity," "recreation," a "learning opportunity" and a "special nature experience."

Health was also linked to the notion of safety. As one respondent put it, "Parents could [sic] feel secure if their children are on the Waterfront Trail. Cycling the trail or walking the trail or looking for frogs in the wetlands are all good, healthy activities that aren't going to get them into trouble."

The notion of place as a safe place to play while providing for a "special nature experience" was also reflected in our interactions with children themselves, who were invited to visit the trail and reflect upon the meaning of the experience of nature in an urbanized environment. The five classrooms of school children with whom we worked were enrolled in Grades 5–9, in three schools west of the City of Toronto. Two groups were from a private school and the remaining three classes were from public schools. We worked with just over a total of 100 children, with whom we traveled portions of the trail and also engaged in classroom activities. All interactions were organized around two questions: (1) Focusing on our experiences in visiting the trail, how would you define nature in the city? And (2) What *should* nature in the city look like?

The data collected in this portion of the study was extensive and included photos, drawings, essays, poems and taped conversations. We recognized that there was some diversity in responses, particularly in light of the fact that one class specifically was not only situated on a marsh setting but engaged in regular activities outside of the classroom and in the surrounding natural environment, which was itself one of the richest and diverse sections of the trail. As one might expect, those children emerged with some of the more discerning descriptions of place, noting details that other children neglected to see.

Students were asked to photograph what they found to be the most important images from their walks. Eric, a Grade Six student, for instance, photographed a fallen tree, explaining that "the reason that I chose this picture was because I thought it was interesting how you see that something's sort of wrecked and you now see that it's trying to create its little branches and stuff and things are growing on it." His friends, also quite discerning, focused their attention on tiny fish eggs in the water. Other students with fewer opportunities for daily environmental engagement by-passed such subtleties, focusing instead upon typical aesthetic pictures of lakeside vistas. Not surprisingly, students with the most hands-on experience of place were more discerning in their descriptions of place.

Despite the differences and nuances of interpretation amongst the variety of students with whom we worked, a finding emerged that was not insignificant in terms of the topics of health and place which we are collectively addressing. In all cases, when asked to address the question of what nature *should* be in the city, by far the most frequently cited response (top in four classrooms and second in the remaining classroom that we studied) related to the theme of health as *cleanliness.*[3]

Fourteen-year-old Leah noted that "nature in the cities should be clean, not like what it is now. Right now, the water is polluted and there is garbage all over the place. I also think that they should clean up the polluted air too. I think that they should clean up all those things and that's what I think nature in the cities should be." Her friend Jade agreed: "I would like [the city] to be clean and to have pure beauty. I wish we could have a nice walk in the park without the pesticide and the garbage. The poor fish die every day from garbage." Interestingly, she concluded that the devastation extends so far that "the birds also can't eat because of the beer cases!" Another student talked about the item that he had chosen to photograph. "A piece of wood – I picked this because, I don't know, I think it's just dirty, the city's dirty."

In a class discussion, another student recognized the challenges of environmental remediation, noting that "it would be really hard for the city to clean up the lake for the fish to actually be able to live in it without sort of being sick or something" and so he recommended building "giant ponds and then there'll be healthy fish and you have to pay $2 for every one you catch, and it doesn't matter what size!"

While some of the main concerns amongst these children certainly did relate to the need for environmental remediation and clean-ups, their stories included another dimension of healthy environments as "clean." In some cases, 2/3 of the class described their notion of a clean environment in terms of "kept up" and manicured places. "I think that they should plant more flowers in the park," said one nine year old. "And I have another thing to add to that. They should put bird feeders around so the birds can get seeds."

Many children saw the merits of providing parks and playgrounds and, once again, these planned areas constituted their vision of what nature in the city "ought" to be. "The government should plant more trees and flowers to make more good looking parks," reflected ten year old Adis. Jasmin agreed with this sentiment: "I think that they could cut the grass in the parks and more people would enjoy going to the parks." Alex suggested that "they should plant more trees and a lot more benches for people to sit and chill with their friends, and fountains because when you are really thirsty,

3 The class that did not predominantly list a "cleaner" environment as the ultimate moral goal, suggested instead that nature in the city should promote exercise and recreational activity.

you do need that refreshment." Ali believed that "the parks should be clean and the water should be filtered, so we can drink water from the lake if we are thirsty."

While the children's perceptions of the value of environmental remediation are, to some extent laudatory, there is also a problematic dimension in their interpretation of clean and healthy environments as manicured and controlled. In the true spirit of master planning, we are told in the class discussion that "they should make two lane paths for people going one way and people going the other way." The classroom conversations continue as one student describes his picture of nature in the city: "It's a picture of my idea, and it has trees, a playground, it has ducks and water. There's pathways, there's places to sit on. There's a fountain and there's a place to sit under, so that sun won't get you, like a shelter. And there's a garbage place and there's grass all over. And a sidewalk to ride your bike or skateboard on it. That's all."

The stories of clean, manicured places described as "natural" abound in our data set. This is not to say that other images of what I will call "messy engagement" do not occur, as students are invited to jump into creeks and explore the water firsthand. "A little later, we walked in the water," Chris B. tells us. "Some of us caught fish in our bare feet but I didn't catch any fish. We saw fish eggs that were yellow. Some of us got wet and dirty because the fish splashed us with their tails. I had a lot of fun and I hope we can do again."

Nevertheless, those spontaneous moments, precious as they were to some students, did not take away from the fact that, overall, the students' prescriptions for bringing nature into the city generally evolved around a notion of cleanliness that was not unproblematic. A safe and healthy environment was seen to be the equivalent of a "clean" environment.

Needless to say, few of us would recommend introducing bacterial or toxic elements into a child's environment. In many ways, health requires cleanliness and it is encouraging to hear the children point to their legitimate concerns over litter.

At the same time, are there some potential problems in equating healthy natural places with "cleanliness?" Might not such notions of cleanliness sometimes equate to sterility? Overly engineered, neatly ordered places, while often physically devoid of dirt, can be aesthetically and emotionally "disinfected" and barren as well.

In his book, *An Ontology of Trash*, Greg Kennedy makes the point that the concept of a sanitizing, throw-away society does not come without some loss (2007). It is not simply a matter of recognizing the wastefulness of building a disposable society but rather, that the concepts of cleanliness and order might potentially deny the reality of life as more rich, more diverse and more "messy" than manicured notions of cleanliness and health will allow.

Others speak about western conceptions of dirt and soil as the source of alienation and lack of immediate connectedness from the earth (Benammar 2000). High rises that take us even further away from the insects, the worms and the "dirty" soil sometimes help to build a conception of place that is sterile, antiseptic and divorced from the messy engagement and the mystery of natural settings.

Places of mystery that engage our imaginations and creative wonder rarely come in the form of neatly planned subdivisions. This is why the results of our study, which show that pre-teenage children already have a manicured vision of environmental health and cleanliness, are potentially disturbing. The children's visions of nature in the city rarely included wild places – those muddy fields and forests which many of us frequented in our own childhood and which prompted us to engage with the wonders and the dangers of the natural world as part of our daily education (For similar conclusions, see Louv 2005). These findings give us pause for thought, as we seek to better understand and develop policies and programs in the areas of education and environment. In the final section of this chapter, I suggest that these sorts of qualitative studies may have a significant role to play in supplementing quantitative data that, ultimately, helps to inform wise environmental decision making in Canada.

Concluding Remarks: Impacting upon Environmental Policy and Practice

There are a number of conclusions that I wish to draw in this chapter. First of all, describing the meaning of such terms as "sense of place" or "health" is often challenging, precisely because those terms refer to phenomena that exceed reductionist parameters. Indeed, inasmuch as these concepts constitute the very condition of the possibility of the appearance of particular *aspects* of place or *facets* of health, they are fundamentally ontological in nature – meaning that defining them in purely calculative terms misses out on their very essence. The mysterious sense of belonging that we might have to a childhood home or that wondrous exhilaration that we experience when we feel when we are in the presence of a rich and healthy old growth forest are not, typically, moments that can be quantified in scientific terms.

To that extent, interpreting such moments requires methodologies that are innovative and non-quantificational in their approach – which brings us to our second conclusion. Qualitative research methods offer tools for investigating human experience of place and good health in uniquely interesting and practical ways. For instance, our own study of the Lake Ontario Waterfront Trail elicited notions of healthy places as diverse, safe – and, in the eyes of children, as "clean."

While in many ways laudable, this latter finding also presents us with an opportunity to carefully review policy implications that would equate "good places" with "clean places." Might the children's vision of the ideal nature trail as manicured and engineered reflect something essential that is missing along this trail? The City of Toronto is actively developing its own waterfront and, in the process, continues to support a passion for condominium construction and landscape uniformity. Even the so-called "motel strip" which originally was to be a landmark development symbolizing the historical preservation of place, presents itself today as a standardized, antiseptic collection of high rises, stacked townhomes and mowed lawns.

Where are the places that meander, that invite curiosity and lingering? Where are the places for spontaneity, the reminders of the grace of a natural world that, ultimately, exceeds our calculative grasp? Without compromising public safety, are there ways in which to introduce places of complexity, of mystery, of imagination to our waterfront – places that are not simply uniformly engineered environments?

These kinds of questions raise policy issues that may not be adequately considered by city planners. If our children see nature through a sanitizing lens already at the age of nine, is this reason enough to rethink our policies on waterfront development and ensure that they incorporate greater environmental complexity?

Many people prefer to envision the relationship between scientific study and policy development as a linear process, where policy is built upon objective, quantitative scientific foundations. In actual fact, as Sarewitz and others have shown, the relationship between science and policy is far more convoluted and iterative, with political pressures influencing policy makers, who rely upon studies that, due to the complexity of the problems under study, rest on scientific uncertainty (2000).

Within this process, one hopes that policy development is based on broad-based consultation, both with scientists as well as with social scientists, humanists and the broader public. There is a new role here for qualitative research studies that explore, differently than traditional quantitative surveys or statistical analyses, the human dimensions of environmental decision making. In fact, there is a new role for audiences of all ages, including children. Including their voices could itself mean throwing the public consultation process more widely to hear different generational views.

Qualitative research studies such as ours certainly can raise questions that may not otherwise arise. In addition, they may reveal opportunities for further quantitative study: in this case, for the purposes of educational curricula, it might be useful to dwell more in-depth in a quantitative survey of children's perceptions on a larger scale.

Policy makers are already recognizing that, in the face of scientific uncertainty, judgment calls and public consultation procedures are providing benefits to decision making. In the end, one must not exclude holistic, qualitative findings from these conversations. Not only will the public consultation process be more inclusive in these circumstances but new insights will likely be revealed that otherwise may have been missed.

References

Barrett, W. (1958), *Irrational Man* (New York: Doubleday).

Benammar, K. (2000), 'Sacred Earth', in R. Frodeman (ed.) *Earth Matters: The Earth Sciences, Philosophy and the Claims of Community* (New Jersey: Prentice-Hall).

Berry, W. (1992), *Fidelity: Five Stories* (New York: Pantheon Books-Random House).

Burke, T.A. (1996), 'Back to the Future: Rediscovering the Role of Public Health in Environmental Decision Making', in R.C. Cothern (ed.) *Handbook for Environmental Risk Decision Making: Values, Perceptions and Ethics* (Boca Raton, London, New York, Washington: Lewis Publishers).

Callicott, B. (1997), 'Fallacious Fallacies and Nonsolutions: Comment on Kristin Shrader-Frechette's 'Ecological Risk Assessment and Ecosystem Health: Fallacies and Solutions', in *Ecosystem Health* 3(3) September (Blackwell Science Inc.).

Casey, E. (1993), *Getting Back into Place: Toward a Renewed Understanding of the Place-World* (Bloomington and Indianapolis: Indiana State University Press).

Clay, G. (1983), 'Sense and Nonsense of Place', *Landscape Architecture* 32: 110–113.

Crabtree, B.F. and Miller, W.L. (1999), *Doing Qualitative Research*, 2nd Edition. (Thousand Oaks: Sage Publications).

Cresswell, J.W. (1998), *Qualitative Inquiry and Research Design: Choosing Among Five Traditions* (Thousand Oaks: Sage Publications).

Descartes, R. (1968), *Discourse on Method and Meditations* (translated by F.E. Sutcliffe) (Middlesex: Penguin Books).

Gadamer, H-G. (1996), *The Enigma of Health: The Art of Healing in a Scientific Age* (translated by J. Gaiger and N. Walker) (Stanford: Stanford University Press).

Heidegger, M. (1962), *Being and Time* (translated by J. Macquarrie and E. Robinson) (New York: Blackwell Publishing).

—— (1966), *Discourse on Thinking* (New York: Harper and Row Publishers).

—— (1971), *Poetry, Language, Thought* (translated by A. Hofstadter) (New York: Harper Colophon Books).

Jager, B. (1985), 'Body, House and City', in D. Seamon and R. Mugerauer (eds) *Dwelling, Place and Environment* (Dordrecht/Boston/Lancaster: Martinus Nijhoff Publishers).

Kennedy, G. (2007), *An Ontology of Trash: The Disposable and Its Problematic Nature* (Albany: State University of New York Press).

Louv, R. (2005), *Last Child in the Woods: Saving our Children from Nature Deficit Disorder* (New York: Workman Publishing).

Malpas, J.E. (1999), *Place and Experience: A Philosophical Topography* (Cambridge: Cambridge University Press).

Merleau-Ponty, M. (1968), *The Visible and the Invisible* (Evanston: Northwestern University Press).

National Research Council (2005), *Decision Making for the Environment: Social and Behavioural Science Research Priorities* (Washington: The National Academies Press).

Norberg-Schulz, C. (1985), *The Concept of Dwelling* (New York: Rizzoli).

Rapoport, A. (1994), 'A Critical Look at the Concept of 'Place', *The National Geographic* Journal of India: 40.

Rapport, D.J. (1997), 'What is Ecosystem Health?', *Ecodecision* Winter: 23, 45–47.

Relph, E. (1976), *Place and Placelessness* (London: Pion).

Sarewitz, D. (2000), 'Science and Environmental Policy', in R. Frodeman (ed.) *Earth Matters: The Earth Sciences, Philosophy and the Claims of Community* (New Jersey: Prentice-Hall).

Seamon, D. and Mugerauer, R. (eds) (1989), *Dwelling, Place and Environment* (New York: Columbia University Press).

Stefãnovic, I.L. (2000), *Safeguarding Our Common Future: Rethinking Sustainable Development* (New York: State University of New York Press).

Chapter 5

Qualitative Approaches in the Investigation of Sense of Place and Health Relations

John Eyles

Introduction

This chapter will examine the use of qualitative approaches for sense of place and health studies. The chapter is also located in the use of qualitative research and the use of qualitative methods in the social sciences and increasingly the humanities. It is also located in the academic arena, and therefore excluding, the work of journalism and social commentators in the eighteenth and nineteenth centuries. Qualitative approaches may be dated from the early twentieth century, especially the work of Malinowski (1922) with his attempt "to grasp the native's point of view, his relation to life, to realize his vision of his world". This attempt to capture the world of others – despite limitations and challenges – remains central to the project of qualitative research. It remains even as qualitative researchers moved to explore and understand complex, often urban societies and frequently their own society. Begun systematically by the Chicago School of Urban Sociology in the 1920s and 1930s, these ethnographers (see Hannerz 1980) used interviews, key informants, statistics and documents to tell the stories of specific social groups. This was taken on by others to study communities or localities throughout the U.S. and Britain in the 1940s and 1950s. Later, attention would be paid to specific environments – factories, schools, and hospitals, geared towards an understanding of meanings and experiences as seen by those in such environments or situations.

Such investigative approaches came somewhat later to place studies as represented by geographical researchers. Ley (1974) utilized interviews, a survey and archival documents to paint a picture of parts of inner Philadelphia, emphasizing geographical exploration as a key starting point. And it was not perhaps until the early 1980s that health geography began utilizing qualitative approaches. Cornwell (1984) carried out a series of in-depth interviews with residents of Bethnal Green, east London, to discover their experiences of health and illness and their connection to economic well-being, while Donovan (1986) used a similar approach – overlaid with an analysis of race, class and employment in Britain – to explore experiences of health, illness and health care among Londoners of Afro-Caribbean and Asian descent. From these beginnings, there has been a massive growth in the use of qualitative approaches in health geography and health research more generally, some of which pertain to senses of place and health. This chapter will now go on to examine this trend as well as some of the challenges identified with such research. Examples of qualitative approaches in health research will be used as illustrations. In these examples, the different techniques of qualitative research will be highlighted. An in-depth analysis of these techniques, which encompass observation (including participant observation), talking to and asking people questions (interviewing), using ways to capture the world of research subjects through such media as film, poems, texts, photos, and a combination of the above, will not, however, be the main focus of this chapter.

The Range of Qualitative Methods for Health – Sense of Place Studies

Dyck (1999) has written cogently on why qualitative methods have grown in importance in health geography. The critique of biomedicine is now well-known (Eyles and Woods 1983; Mohan 1989) and need not be repeated. But while quantitative approaches remain vital for looking at illness and disease, the linking of environment and health and to investigate access to care, interest has shifted to experiences of health, illness and health care and the structural forces that affect individuals. The use of qualitative methods furthermore helped align place studies with the cultural turn in the social sciences and humanities. An early example is Kearns' (1991) study of clinics in Hokianga, New Zealand, using interviews and participant observation. In this paper, Kearns comments that the selection of the actual qualitative approach adopted is personal and pragmatic – a notion that still underlies much qualitative research in the social sciences in general. Indeed, Dyck (1999) points to the array of methods available – in-depth interviews, focus groups, participant observation, oral histories, journal-keeping, autobiography, photographs and textual analysis. Through text, all kinds of documents are amenable to qualitative research and analysis. Dyck does not comment on any method preference, noting that these methods singly or in combination enable an explication of the complexities of people, places and health. Indeed, she emphasizes their use in conjunction with other methodological and theoretical interests rather than their role in discovery and knowledge creation alone.

In taking this approach, Dyck helps in identifying the broad advantages of place studies' engagement with qualitative research. Thus she argues that the creative use of quantitative (usually epidemiological survey) with qualitative methods (usually focus groups, in-depth interviews and document analysis) in environmental health has led to the revealing of experiences and context with, for example, the understanding and explication of risk. Secondly, Dyck suggests that qualitative methods have enable a decentring of the medical paradigm for understanding people and places with an emphasis on the individual's story or narrative and with how behaviour and concepts of self and care are deeply embedded in place. Indeed, this decentring was an early *raison d'etre* for the use of qualitative methods with health ideas and behaviour emerging from everyday circumstances (see Litva and Eyles 1994). Thirdly, Dyck identifies the desire to use theory, especially in conducting applied research. For example, Gesler (1993) uses cultural theory to better understand the ideas of health and healing around therapeutic landscapes. We may note the central role of identifying senses of place in these reasons.

Dyck's important paper identifies the research strategies in qualitative design and the importance of theory in undertaking such research practice. Similar emphases may be found in Creswell (1998, 2007) who provides a manual for researchers. He points out that "qualitative research is complex, involving fieldwork for prolonged periods of time, collecting words and pictures, analyzing this information inductively while focusing on participant views, and writing about the process using expressive and persuasive language" (1998, 24). Moreover, the research should be framed within and by a tradition of inquiry which provides a stance toward theory and a succinct approach to the research problem at hand. Creswell identifies five traditions – biography, phenomenology, grounded theory, ethnography and case studies – and it is an important comment that no one of these is privileged over the others. His approach provides a useful, if limited, guide as much qualitative research uses several of these traditions simultaneously. Indeed some advocate combining the perspectives. For example, Hansen (2006) suggests that narrative and ethnography are both necessary to obtain understanding of text and context. Much then depends on the research questions in hand.

In summary, the five traditions are widely understood as *narrative* seen as a spoken or written text giving a chronologically connected account of actions or events (Czarniawska 2004) and connected to the ways in which people learn about, explain and organize experience (Wiles et al. 2005). Narrative includes *biography* – a study of an individual and his/her experiences as told to the researcher or found in documents and archives, often in the form of a life history; *phenomenological study* – a description of the meaning of lived experiences about a concept or phenomenon; *grounded theory study* – while emphasizing the meaning of experience, its purpose is to discover theory relating to that situation; *ethnography* – a description and interpretation of a cultural or social group or system; and *case study* – an exploration of a 'bounded system' or case over time through detailed, in-depth data collection involving multiple sources of information rich in context. Creswell goes on to link these traditions to the use of theory, seeing theory as informing an ethnography or phenomenological study but being the product of a grounded theory approach. Theory is used in different ways in narrative and case study. Creswell then, somewhat unproblematically, possibly naively uses these traditions with multiple research strategies – observation, interviews etc. – to illuminate research questions. In fact, we too will suspend the challenges that emerge in the use of qualitative inquiry for the time being to illustrate these approaches in health – place studies and cognate fields. It is, in my view, more useful than illustrating the methods or strategies per se, these being utilized in many different settings and for many different research questions as are statistical sources and surveys.

Qualitative Inquiry in Sense of Place – Health Studies: Some Examples

Narrative

Narrative is perhaps the least well-represented approach in place studies but it has been one where verbal, as well as non-verbal and non-text based strategies have been used. The use of photos, camera and video are present in urban anthropology and educational research. A visual narrative inquiry allows the individual's life to be connected to culture, context and past to make visible its different parts, even that which may be repressed or resisted (see Berger 1980; Bach 2001). Combining photographs and interviews, Scuro (2004) explores the life of an Italian immigrant woman in the U.S. She uses immigrant testimony to enrich our understanding of the struggle with traditional ways and patriarchy and what migration meant for the creation of identity and the establishment of well-being.

More specifically, photovoice (Wang and Burris 1997) has been used to enable people to identify and represent their community through photos and related dialogue. Examples include the portrayal of health and community realities among the homeless (Wang et al. 2000) and for access to care for transsexuals (Hussey 2006). With the telling of the cultural stories of their research subject through interviews, Chan et al. (2006) examine the meaning of health among Hong Kong Chinese adults while Milligan (2005) uses correspondence (i.e. letters, photos and diaries) to look at the experience of social care in New Zealand. Finally, using primarily written sources, Brown and Moon (2004) provide a biography of Jacques May, one of the founders of medical geography. While the purpose of their paper is broader, Brown and Moon demonstrate, through linking May's autobiography to the contexts of French colonial rule and its perceived civilizing function and American medical and military agendas, the importance of biography to understanding any individual in his or her era. Narrative description is thus a diverse and rich approach, permitting the identification of identity and belonging through the individual positioning his/herself through the story (see Kraus 2006) but with the challenge of combining narratives as interpretations of the real world to accumulate knowledge about the world (see Josselson 2006).

Phenomenological Studies

Phenomenological investigations intend to expose the essential characteristics of a concept or phenomenon. Such investigations can be found in place studies and related disciplines but they are seldom related to health. Furthermore, they depart from Creswell's (2007) preference for psychological phenomenology and may be said to have more in common with Schutz's sociological version (see Eyles 1985). Indeed, it is perhaps wise to differentiate between transcendental or descriptive and hermeneutic phenomeology (see Rapport and Wainwright 2006). The former, based on the philosophy of Husserl, describes the way we see and determine meaning in the world. Giorgi (1997, 242) argues that the essence of any phenomenon "is the most invariant meaning for a context. It is the articulation, based on intuition, of a fundamental meaning without which a phenomenon could not present itself as it is".

The latter, based on the philosophy of Heidegger, leads to understanding through interpretation and the continuous re-examination of propositions. As Ricoeur (1974) puts it, we understand and interpret through our relationship with the world in which we are already and always a part. And as Rapport and Wainwright (2006) argue for nursing research, both approaches have significant merit. The same is surely true for health-sense of place investigations.

For both types of phenomenology, experience of phenomena is key – indeed as Relph (1976, 6) puts it, it is "the entire range of experiences through which we all know and make places". This therefore relies much on reflection and inductive argument based on immersion in all information available. A context-rich understanding of a phenomenon may be seen in Philo's (1995) work on the journey to the asylum in nineteenth century England which enabled him to challenge Jarvis' law concerning the spatial relations between medical provision and its clients. We may note the wide-ranging arguments of both Jarvis and Philo to help understand the phenomenon as experienced by "lunatics" – geographical, interventionist, social, economic and moral to distill the meaning of this dimension of the "mad business".

Other examples of phenomenological investigations may be found in design and planning, focusing on the creation and identification of senses of place as yet only implicitly linked to health. Indeed, Aravot (2002) argues that the intensive critique of phenomenology may have had, as an unintended consequence, the denigration of place-making, arguing that place is in fact a phenomenological term and its experience consists of indispensible knowledge about the world and about the condition of human existence in that world. Aravot therefore asserts that phenomenological place-making is a guiding principle. She goes on to cite Merleau-Ponty to see places as "shared echoes within ourselves as human beings, aroused by material and formal attributes" (2002, 209). Place is then not just perceived by sight but by all senses and experience. Arkette (2004) for example has used phenomenology to understand and interpret the soundscapes of cities. Such a view holds great promise for place and health studies as the sounds, smells, touch, tastes and feelings of places can be related to health. In Raban's (1975) terms, the city is plastic – providing opportunities (and constraints) for health-related experiences.

Furthermore, if we follow Creswell (1998, 2007) and see our phenomenological investigation shaped by a particular theoretical stance, further consideration of phenomenological study is possible. Smith and Bugni (2006) infuse their examination of the meaning of architecture with symbolic interactionism with a focus on the meanings assigned to objects and their impact on us (Blumer 1969). Thus architecture reflects and expresses the self, providing places of the soul where physical shapes, forms, spaces and appearances provide a picture of reality that nourish emotion and selves through our experience of them (see Day 1990). Places also condense meanings and values, becoming key dimensions in a system of communication used to articulate social relations (see Lawrence and Low 1990). From this base, Smith and Bugni go on to reflect on the experience of architecture. Many of their examples are relevant to health and well-being as sites convey control and promote (or inhibit) meaning and agency. Place studies have pursued such ideas, primarily through attention to therapeutic landscapes (see Williams 1998; Baer and Gesler 2004). But architecture and sites in general may, if they are cherished, become artifacts that incorporate culture, time, and space for the beholder, helping define self in the world (see Baudrillard and Nouvel 2003). But there is also a need to investigate landscapes of fear (see Tuan 1979; Gold and Revill 2003). For health – sense of place, the rigours of a theory-informed phenomenological investigation of places and health seems an important way forward seen already in health research with Midtgaard et al. (2007) use of phenomenology to underpin the interpretation of narratives with respect to the activities of late-stage cancer patients.

Grounded Theory Studies

While common in nursing (see Carolan et al. 2006), grounded theory is also little used in place studies. Through emphasizing the nature of experience, this approach purposely seeks to generate theory emerging from the research situation. One implicit use may be found in James and Eyles (1999) in their investigation of perceptions of health and environment in contrasting neighbourhoods in Hamilton, Ontario. Using in-depth interviews, James and Eyles identified detailed perceptions about health but more generalized ones about environment, as well as ideas about how health and environment were connected. Issues of concern for self, others and the environment and of control and responsibility were documented. James and Eyles then asked why did these particular perceptions emerge. This question led them to develop a cultural model of health-environment relations "grounded" in their respondents' talk. In this way, they contributed to a theory of mental model development and use based on Fiske and Kinder's (1985) notion of cognitive misers who painfully learn premises, dogma, habits, theories, codes and so on (see also Lane 1962) so that experiences and perceptions are filtered through and into existing and simplified representational models.

Other examples may be found in related disciplines. Thus Agier (2002) examines refugee camps which are supposed to provide safety, protection and the basic necessities of life for displaced peoples. Through the detailed description of life in the Dadaab camps in Kenya, Agier can outline the nature of identity, arguing that not only do the camps create identity but do so by configuring a bricolage of new identities which emphasize specific ethnic particularisms. Agier comes up then with a process of ethnic identity creation and maintenance, especially important as the camps are places of inter-group socialization and have developed a permanence belied by their creation as emergency, and hence short-term, shelters. Furthermore, study of such "permanent" camps may lead to a new theorizing of urbanization and the provision of services to such communities. One further example is taken from a study of pain experiences among Aboriginal women in rural Australia (Fenwick and Stevens 2004). Findings from this study were used to critique existing pain relief strategies and to develop new ways for non-Aboriginal nurses to carry out their work, likely more in keeping with Aboriginal expectations, namely to act as traditional healers do.

Ethnography

More common is ethnography, the description of a social group or system. An ethnography provides a thick description of meanings and experiences within the encompassing context. An early example is provided by Eyles and Donovan's (1986) study of sickness and care in "Mossley Green" in the English west Midlands. Using census material and observation, they begin by describing the physical and social characteristics of the neighbourhood, the backdrop to engage residents through in-depth, semi-structured interviews in talking about health and illness in their everyday lives. The perceptions of health now seem commonplace. There is a desire to see oneself as healthy, to set aside health problems, even quite serious ones, as minor inconveniences that are lived with and through. Furthermore, what makes individuals healthy or not – and it is often that the self is healthy and rises above illnesses whereas others succumb to their ailments – is in part anxiety, stress and worry as well as lifestyle and behaviour. But fate, powerlessness and inevitability are also seen as "causes" of sickness. Such reasons provide for a richer contextualizing of ill-health. Heredity is often noted as important – "cancer runs in the family". Fate and its opposite, luck, are also implicated – "good health is a matter of good luck". So too are structural conditions, such as unemployment, neighbourhood, type of work, physical environment and social and civic issues such as crime and race. These elements are brought together by individuals, Eyles and Donovan argue, to make sense of health and illness in what is often a hard, deprived everyday existence. And making sense is made more difficult in changing circumstances which require a sense of historical existence and continuity.

A more recent example is provided by Singer et al. (2000) in their explication of the social geography of AIDS and hepatitis risk in three U.S. cities, Hartford, New Haven and Springfield. They argue that a single data collection approach cannot provide a thick (i.e. rich and context-dependent) and comparative description of the experiences of injection drug users (IDU) and the risks they bear. Singer et al. use, therefore, six qualitative methods in combination with an epidemiological survey and bioassay procedures to understand access to sterile syringes in these cities, the condition of places used for drug injection and cultural practices around drug use. The six methods were neighbouhood-based IDU focus groups to construct social maps of local equipment acquisition and drug use sites; observation and description of target neighbourhoods; IDU diary keeping and acquisition and use; natural setting observation and interviews through day visits with IDUs; interviews with IDUs to collect syringes for laboratory analysis; and observation and interview during drug injection. The authors argue that together those methods provide the basis for an ethnography – a geographic ethnography – of the importance of local context for IDU risk behaviour, how this context varies between micro-environments and how these factor conspire to promote or inhibit the transmission of HIV and other blood-borne diseases.

Case Study

Perhaps the most common qualitative approach in place studies is the case study, the description of a "bounded" system through multiple data sources. "Case study" is often used as a sub-title in many qualitative studies. But only three examples will be provided, one of a waterborne disease outbreak, the second of access to health care and the last of a small social environment. The first refers to the case of contaminated drinking water in Walkerton, Ontario. Over 2300 people became seriously ill and seven died from exposure to *E. Coli*. In one account, Hrudey et al. (2003) outline other outbreaks of waterborne disease before examining Walkerton by carefully reviewing source, water treatment, water distribution, well monitoring and contaminant response from multiple data sources – geologic reports, water records, public health responses and management activities – to document the multiple failures that led to the outbreak. Perkel (2002) uses media reports and interviews with key informants and residents to chart the causes, context and reality of the outbreak, Snider (2004) examines the aftermath Inquiry itself as a political and legal event to understand the role of government and science in this case and Ali (2004) provides a socio-ecological frame to understand the links between health, place and toxicity. Parr (2005) looks at the meaning of "good water" and community response in Walkerton. An investigation of what Walketon now means practically for residents' health and metaphorically for that of other Ontarians remains to be done.

Secondly, Wellstood et al. (2006) use a case study of primary care provision in Hamilton, Ontario to examine access to care. Using in-depth interviews of residents in two neighbourhoods, they investigated experiences of accessing and receiving care from primary care health professionals. From these interviews, they were able to identify two sets of factors shaping their experiences – individual and system. System barriers included wait times, actual location of the doctor's office, limited hours of operation and lack of services offered by the doctor. Individual constraints include work and family responsibilities. The authors conclude that system barriers – including location and place – limit "reasonable" access, a principle of health care provision in Canada. While perceptions of access to services and of services themselves have been investigated as in Scoular et al's (2001) study of genitourinary services, senses and experiences of the places where services are offered might be rewarding, especially with respect to the uptake of service opportunities. Finally, Yodanis (2006) examines the social interactions and well-being in the micro-environment of the coffee shop. Through an ethnomethodological frame of interaction and participant observation, she explored leisure, family and class to understand how self-identified social position and social worth shaped interaction exchanges. Furthermore Yodanis lived in the town, engaging in community life as much as possible. Indeed, this study is a rich melange of qualitative approaches but what is reported is the case study of the coffee shop. It is a good example with which to end this section, as she points to the importance of qualitative inquiry in social research, approvingly citing Zussman (2004, 352): "qualitative sociology works best when it addresses people in places" – a fine commentary on the utility of such approaches for health – sense of place studies.

Challenges to Qualitative Inquiry and a Way Forward

The examples chosen for the previous section demonstrate the richness of qualitative inquiry. They reveal the many different data sources and research strategies used, although they have tended to emphasize interview rather than documentary data. The importance of discourse analysis in health and place should not be overlooked. Snider (2007) provides an introduction to such an approach for health research. Furthermore, Allen and Hardin (2001) provide a wide-ranging discussion of discourse analysis of various health problems, including eating disorders and mental health while Iannantuano and Eyles (1999) examine the changing nature of policy discourse for setting the agenda for Great Lakes remediation. A potentially useful example is provided by Keil and Debanné (2005). They examine different levels of environmental discourse in the use of place and politics for environmental change. Senses of place are not therefore uniform or universal and this in itself will affect quality of life, health and policy responses.

Yet most qualitative work in health geography consists of interviews (with some document and census analysis) presented as case studies. Put differently, Herbert (2000) recently estimated that only 3–5 percent of journal articles are ethnographic. Why should that be? There is no easy answer to this question. Herbert suggests methodological weakness in ethnography itself, although all qualitative approaches are subject to interpretive problems. Crang (2002) is certainly right to suggest that funding is problematic for such immersed studies. But other issues can be invoked. Career development in the modern academy demands a steady flow of research outputs. Case studies of bounded systems provide these more easily than the other approaches. Furthermore, for health geography, there may be a tension in its qualitative research between the reflective (and therefore often long-term) nature of such work and the pressures of its cognate scientific disciplines – especially medicine and epidemiology – which want to provide responses to questions (solutions to problems) in a reasonably limited period of time.

These tensions also direct attention to philosophical and methodological as well as the identified practical ones. It may be argued that there are philosophical tensions between such an approach in transcendental phenomenology and those relying on the discovery of the empirical – e.g. case study – with a possibility of a retreat to reductionism. The question of role of theory in qualitative inquiry does not lend itself to a simple answer. Virtually all qualitative researchers regard theory as an integral part of their projects. Theory helps establish research questions. Furthermore, it may be derived from empirical investigation, allowing informed commentary on how the world works or why health is configured in particular ways for people in places. Avis (2003) has, however, challenged the utility of methodological theory, suggesting that positivism was a doomed attempt to derive the empirical foundations of knowledge by separating theory and evidence. Qualitative techniques theory may therefore hinder critical reflection and may cloud the relationship between theory and evidence. He suggests that phenomenology, grounded theory and ethnography may obscure the basis of knowledge claims. This is indeed a challenge and there must be a scientific reason for selecting an approach to qualitative inquiry, deriving primarily from the existing state of knowledge and plausible theories of how meaning and experiences may be produced.

Such a conclusion is supported by Johnson (1999) who is concerned about utilizing ideas from the margins of science with qualitative methods being somewhat "mystical" in explication. Certainly, some in place studies have emphasized the important of critical insight and the reflective and reflexive nature of the research enterprise (see Bailey et al. 1999). This has led to a concern about verificationism and the danger of qualitative empiricism (see Crang 2002). But rigour and clarity around objectives, research questions, the use of theory and the nature of interpretation seem vital. In a fascinating paper on limiting cases, Paley (2005) points to the importance of identifying and eliminating all "error" using delusional patients as a study population: what the subject says is itself open to multiple interpretation and "error". So given these challenges to theory and method for interpretation, how do we proceed?

Key in proceeding must involve clarity concerning definitions, procedures, interpretation and verification. Baxter and Eyles (1997) proposed a rigorous approach to interview data, using Lincoln and Guba's categories of credibility (authenticity of the account), transferability (intelligibility), dependability (adequacy and plausibility of a full account) and confirmability (through audit, member-checking and so on). There seems to be wide agreement from many disciplines using qualitative methods that such rigour is needed from project conception to interpretation and reporting (see Koch 1994). Charmaz (2004) in fact revisits the premises and principles of qualitative research. Her arguments are worth exploring. With respect to premises, she suggests a good qualitative analysis depends on entering what is studied to obtain a deep understanding of meanings and experiences. Yet significant meanings may be unstated or silenced, so that actions may make taken-for-granted meanings visible. Further, the question asked (and theories used) determine the answers obtained, which will themselves be variable but still "truthful". With respect to principles, she argues that intimate familiarity with the phenomenon is necessary; phenomena should be studied at the individual and social levels; extant theory is a beginning not the end point; scientific interpretations (constructions, typifications) complement the studied phenomena; and respect for subjects supercedes research objectives. These are not easy to put into practice. But as Rapport et al. (2005) note, there is an ethical imperative to ask questions using a qualitative approach to illuminate meaning and experiences in times of illness and treatment. Qualitative techniques – interviews, archives, documents, observations, photos, video and so on – enrich the subject-matter of senses of place and health to deepen our understanding of health and illness in places, as might the various approaches described in this chapter. We should recall that none was privileged. Indeed some have argued that there are no such things as "pure" qualitative approaches (see Johnson et al. 2001), pointing the importance of revealing that which may be socially and intellectually embedded in our approaches and tools (see also Bourdieu 1992). Thus the case for their utility then seems made. What is now needed are further studies using their techniques and methodologies to discover the nuanced shapes of the relations between senses of place and health: a need discussed in the concluding chapter of this book.

Acknowledgement

The author wishes to thank workshop participants in Canada and South Africa for their helpful discussions on method. This chapter extends an earlier French-language version.

References

Agier, M. (2002), 'Between war and the city', *Ethnography* 3: 317–41.

Ali, S. (2004), 'A socio-ecological autopsy of E.coli 0157: H7 outbreak in Walkerton, Ontario, Canada', *Social Science and Medicine* 58: 2601–2612.

Allen, D. and Hardin, P. (2001), 'Discourse analysis and the epidemiology of meaning', *Nursing Philosophy* 2: 163–76.

Aravot, I. (2002), 'Back to phenomenological placemaking', *Journal of Urban Design* 7: 201–212.

Arkette, S. (2004), 'Sounds like city', *Theory, Culture and Society* 21: 159–68.

Avis, M. (2003), 'Do we need methodological theory to do qualitative research', *Qualitative Health Research* 13: 995–1004.

Baer, L. and Gesler, W. (2004), 'Reconsidering the concept of therapeutic landscapes in J.D. Salinger's *The Catcher in the Rye*', *Area* 36: 404–13.

Bach, H. (2001), 'The place of the photograph in visual research', *Afterimage* (accessed May 29, 2007) http://findarticles.com/p/articles/mi_n2479/is_3_29/ai_80757500.

Bailey, C. et al. (1999), 'Evaluating qualitative research, dealing with the tension between science and creativity', *Area* 31: 169–83.

Baudrillard, J. and Nouvel, J. (2003), *The Singular Objects of Architecture* (Minneapolis: Univ. Minn. Press).

Baxter, J. and Eyles, J. (1997), 'Evaluating qualitative research in human geography', *Institute of British Geographers* 22: 505–25.

Berger, J. (1980), *About Looking* (London: Writers and Readers Press).

Blumer, H. (1969), *Symbolic Interactionism* (Englewood Cliffs: Prentice Hall).

Bourdieu, P. (1992), 'The practice of reflexive sociology', in P. Bourdieu and L. Wacquant (eds) *Invitation to Reflexive Sociology* (Cambridge: Polity Press).

Brown, T. and Moon, G. (2004), 'From Siam to New York: Jacques May and the 'foundation' of medical geography', *Journal of Historical Geography* 30: 747–63.

Carolan, M. et al. (2006), 'Writing place,' *Nursing Inquiry* 13: 203–19.

Chan, E. et al. (2006), 'A narrative inquiry into the Hong Kong Chinese adults' concepts of health through their cultural stories', *International Journal of Nursing Studies* 43: 301–9.

Charmaz, K. (2004), 'Premises, principles and practices in qualitative research', *Qualitative Health Research* 14: 976–93.

Cornwell, J. (1984), *Hard-earned Lives* (London: Tavistock).

Creswell, J. (1998), *Qualitative Inquiry and Research Design* (Beverly Hills: Sage).

—— (2007), *Qualitative Inquiry and Research Design* 2nd edition (Beverly Hills: Sage).

Crang, M. (2002), 'Qualitative methods: the new orthodoxy?', *Progress in Human Geography* 26: 647–55.

Czarniawska, B. (2004), *Narratives in Social Science Research* (Thousand Oaks: Sage).

Day, C. (1990), *Place of the Soul* (Wellingborough: Aquarian).

Dyck, I. (1999), 'Using qualitative methods in medical geography', *Professional Geographer* 51: 243–53.

Donovan, J. (1986), *We Don't Buy Sickness, It Just Comes* (Aldershot: Gower).

Eyles, J. (1985), *Senses of Place* (Warrington: Silverbrook Press).

Eyles, J. and Donovan, J. (1986), 'Making sense of sickness and care', *Transactions of the Institute of British Geogeographers* 11: 415–27.

Eyles, J. and Woods, K. (1983), *Social Geography of Medicine and Health* (London: Croom Helm).

Fenwick, C. and Stevens, J. (2004), 'Post-operative pain experiences of central Australian aboriginal woman', *Australian Journal of Rural Health* 12: 22–27.

Fiske, S. and Kinder, D. (1981), 'Involvement, expertise and schema use', in N. Cantor and J. Kihlstrom (eds) *Personality Cognition and Social Interaction* (Hillsdale: Erlbaum).

Gesler, W. (1993), 'Therapeutic landscapes', *Environment and Planning D* 11: 171–89.

Giorgi, A. (1997), 'The theory, practice and evaluation of the phenomenological method as a qualitative research procedure', *Journal of Phenomenological Psychology* 28: 235–260.

Gold, J. and Revill, G. (2003), 'Exploring landscapes of fear', *Capital and Class* 80: 27–50.

Hannerz, U. (1980), *Exploring the City* (New York: Columbia UP).

Hansen, H. (2006), 'The ethnonarrative approach', *Human Relations* 59: 1049–1075.

Herbert, S. (2000), 'For ethnography', *Progress in Human Geography* 24: 550–568.

Hrudey, S. et al. (2003), 'A fatal waterborne disease epidemic in Walkerton, Ontario', *Water Science and Technology* 47: 7–14.

Hussey, W. (2006), 'Slivers of the journey', *Journal of Homosexuality* 51: 129–58.

Iannantuano, A. and Eyles, J. (1999), 'Environmental health policy', *Environmental Management* 26: 385–392.

James, J. and Eyles, J. (1999), 'Perceiving and representing both health and environment', *Qualitative Health Research* 9: 86–104.

Johnson, M. (1999), 'Observations on positivism and pseudoscience in qualitative nursing research', *Journal of Advanced Nursing* 30: 67–73.

Johnson, M. et al. (2001), 'Arguments for British pluralism in qualitative health research', *Journal of Advanced Nursing* 33: 243–249.

Josselson, R. (2006), 'Narrative research and the challenge of accumulating knowledge', *Narrative Inquiry* 16: 3–10.

Kearns, R. (1991), 'The place of health in the health of place', *Social Science and Medicine* 33: 519–530.

Keil, R. and Debbané, A.-M. (2005), 'Scaling discourse analysis', *Journal of Environmental Policy and Planning* 7: 257–276.

Kraus, W. (2006), 'The narrative negotiation of identity and belonging', *Narrative Inquiry* 16: 103–111.

Koch, T. (1994), 'Establishing rigour in qualitative research', *Journal of Advanced Nursing* 19: 976–986.

Lane, R. (1962), *Political Ideology* (New York: Free Press).

Lawrence, D. and Low, S. (1990), 'The built environment and spatial form', *Annual Review of Anthropology* 19: 453–505.

Ley, D. (1974), *The Black Inner City as Frontier Outpost* (Washington: AAG).

Litva, A. and Eyles, J. (1994), 'Health or healthy?' *Social Science and Medicine* 39: 1083–1091.

Malinowski, B. (1922), *Argonauts of the Western Pacific* (London: RKP).

Midtgaard, J. et al. (2007), 'Regaining a sense of agency and shared self-reliance', *Scandinavian Journal of Psychology* 48: 181–90.

Milligan, C. (2005), 'Place narrative correspondence in the geographer's toolbox', *NZ Geographer* 61: 213–224.

Mohan, J. (1989), 'Medical geography', *Antipode* 21: 166–177.

Paley, J. (2005), 'Error and objectivity: cognitive illusions and qualitative research', *Nursing Philosophy* 6: 196–209.

Parr, J. (2005), 'Local water diversity known', *Environment and Planning D* 23: 251–271.

Perkel, C. (2002), *Well of Lies* (Toronto: McClelland Stewart).

Philo, C. (1995), 'Journey to the asylum', *Journal of Historical Geography* 21: 148–168.

Relph, E. (1976), *Place and Placelessness* (London: Pion).

Raban, J. (1975), *Soft City* (London: Fontana).

Rapport, F. et al. (2005), 'Of the edgelands: broadening the scope of qualitative methodology', *Medical Humanities* 31: 37–41.

Rapport, F. and Wainwright, P. (2006), 'Phenomenology as a paradigm of movement', *Nursing Inquiry* 13(3): 228–236.

Ricoeur, P. (1974), *The Conflict of Interpretations* (Evanston: Northwestern University Press).

Scoular, A. et al. (2001), 'That sort of place … where filthy men go', *Sexually Transmitted Infections* 77: 340–343.

Scuro, J. (2004), 'Exploring personal history', *Oral History Review* 31: 1–30.

Singer, M. et al. (2000), 'The social geography of AIDS and hepatitis risk', *American Journal of Public Health* 90: 1049–1056.

Smith J. (2007), 'Critical discourse analysis for nursing research', *Nursing Inquiry* 14: 60–70.

Smith, R. and Bugni, V. (2006), 'Symbolic interaction theory and architecture', *Symbolic Interaction* 29: 123–155.

Snider, L. (2004), 'Resisting neo-liberalism; the poisoned water disaster in Walkerton, Ontario', *Social and Legal Studies* 13: 265–289.

Tuan, Y.F. (1979), *Landscapes of Fear* (Oxford: Blackwell).

Wang, C. and Burris, M. (1997), 'Photovoice', *Health Education and Behavior* 24: 369–387.

Wang, C. et al. (2000), 'Who knows the streets as well as the homeless?', *Health Promotion and Practice* 1: 81–89.

Wellstood, K. et al. (2006), 'Reasonable access to primary care', *Health and Place* 12: 121–130.

Wiles, J. et al. (2005), 'Narrative analysis as a strategy for understanding interview talk in geographical research', *Area* 37: 89–99.

Williams, A. (1998), 'Therapeutic landscapes in holistic medicine', *Social Science and Medicine* 46: 1193–1203.

Yodanis, C. (2006), 'A place in town', *Journal of Contemporary Ethnography* 35: 341–66.

Zussman, R. (2004), 'People in places', *Qualitative Sociology* 27: 351–63.

Chapter 6

Developing a Psychometric Scale for Measuring Sense of Place and Health: An Application of Facet Design

Allison Williams, Christine Heidebrecht, Lily DeMiglio, John Eyles, David Streiner and Bruce Newbold

The ways in which environmental factors influence health have been studied extensively in geography, anthropology, psychology, epidemiology and a wide range of other disciplines (Williams 1998; Mitchell et al. 2000; Williams 2002; Frumpkin 2003; Altschuler et al. 2004; Fletcher 2006). Environment can be examined on many levels, from the micro-environment of the home or community through to the national or global sphere. The impact that place can play as a health care setting – that is, the care environment – for individuals who are unwell, as well as their caregivers, has been examined, as well as the health effects that a home environment can have for both healthy and unwell individuals (Ulrich 1984; Dunn 2002; Williams 2002; Rioux 2005). Looking more broadly than an individual home or other physical structure, neighbourhood characteristics and attributes have also been examined with respect to the role they play in affecting health (Mitchell et al. 2000; Luginaah et al. 2001; Dunn 2002; Wells 2005; Wilson 2007).

Restorative environments literature points to the potential of environments, whether natural or built, to promote healing and otherwise other positive effects (Korpela and Hartig 1996; Hartig and Staats 2003). As indicated above, there is a range of dimensions that can characterize environment: home, neighbourhood, municipality, province and nation. With respect to restorative spaces, micro levels are often discussed: home, a neighbourhood, a localized area within a municipality such as a park, beach, restaurant, or plaza (Korpela and Hartig 1996; Herzog et al. 2003).

What remains largely absent from the literature, however, is the relationship between health and the place-based construct *sense of place*. Sense of place is a construct that embodies the meaning and significance that individuals attribute to the area in which they live. There is some discord across disciplines regarding the components that constitute sense of place, but rootedness, belonging, place attachment, and meaningfulness are elements that have been incorporated into notions and interpretations of sense of place (Shamai 1991; Hay 1998; Manzo 2003; Kianicka et al. 2006). A significant body of theoretical and empirical studies describes sense of place as an outcome of interconnected psychological, social and environmental processes in relation to physical place(s). Sense of place has been examined, particularly in human geography, in terms of both the character intrinsic to a place as a localized, bounded and material entity, and the sentiments of attachment/detachment that humans experience and express in relation to specific places (Relph 1976; Tuan 1974; Gesler 1992; Cresswell 1996; McDowell 1997; Oakes 1997).

As a result of an exhaustive theoretical/conceptual and empirical review of the literature specific to sense of place, the core elements of the construct are found to include: rootedness, belonging, place identity, meaningfulness, place satisfaction and emotional attachment. While these comprise psycho-social constructs, many authors have identified the importance of the physical environment as another core element of sense of place (Eyles 1985; Clark and Stein 2003; Pretty et al. 2003; Stedman 2003a; 2003b; Yung et al. 2003). Place-based research contributes to better understanding of environment as a health determinant (in contrast to approaches that do not consider environmental factors) and may be applied to solutions for improved health, such as enhancing the health promoting qualities of places (Macintyre et al. 1993).

Although we have some evidence that sense of place contributes to community-level pathways and processes that influence health (Warin et al. 2000; Theodori 2001; Kearns et al. 2000; Meegan and Mitchell 2001; Healey 1998), the physiological and biochemical mechanisms of which are not well understood, little operationalization of the sense of place construct has been done, precluding the analytical examination of how it relates to health outcomes.

It is hypothesized that sense of place is an interdependent factor in a variety of determinants of health, health practices, health outcomes and quality of life. If sense of place is found to be an important link in the health pathway(s) of population trends (and this would involve sophisticated modeling and reliable data in future research), population health researchers would have an important new insight on the mediating or moderating factors in determining health outcomes.

The objective of this study was to develop and validate a measurement instrument for sense of place. This chapter focuses particularly on the facet design process used for item construction within the tool development process. The literature review, focus group, and qualitative analysis phases will first be outlined, followed by a detailed description of the item development process using the resultant data. The evaluative steps taken to determine which of the created items should be included in the measurement tool will be illustrated and lastly, the analyses that will be carried out to assess the tool's validity and reliability are described.

Operationalizing *Sense of Place*; The Tool Development Process

Figure 6.1 illustrates the complete tool development process. The facet design process is the focus of this chapter, and is identified with a grey box within the figure. Each component of the process is described below.

The literature review, together with focus group data, were fundamental components of item construction; these two sources of data facilitated the generation of themes for the facet design process. These steps in the research project are described in detail in the following section. Approval for the collection of focus group data was granted by McMaster's Research Ethics Board.

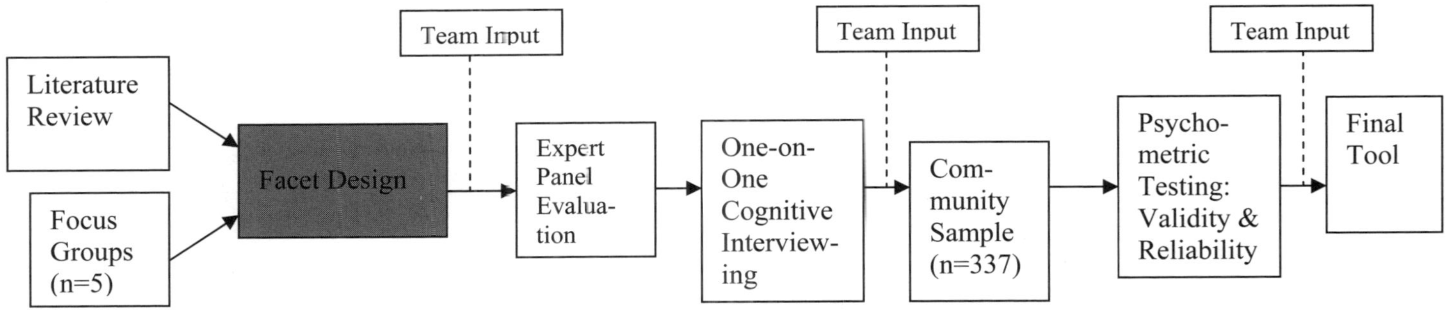

Figure 6.1 Operationalizing *Sense of Place*

Item Construction

Item construction for the measurement tool was informed by both a literature review and qualitative data from focus group discussions. Drawing on data from these distinct sources provided a rich and comprehensive base of information that was used to determine appropriate sense of place themes required for facet design.

The literature was searched for studies and other reports on elements of sense of place. The terms used to search social science and sciences databases, such as Social Sciences Abstracts, Applied Social Sciences Index and Abstracts, GEOBASE, Science Direct, JSTOR, and PsycINFO, included: *place, sense of place, rootedness, place identity, neighbourhood* and *topophilia.* Because sense of place is so broad in nature, the search was not limited to geography databases. Journals with a specific place-related focus were also searched using these terms. Once citations had been obtained, each of their reference lists were examined for additional relevant publications. Surveys pertaining to sense of place and related components were reviewed and relevant items collected. Themes that emerged from other, non-survey literature on sense of place were also documented.

In order to complement the academic literature and to gain an understanding of how individuals interpret and speak about sense of place, focus groups were conducted. Five focus groups were carried out in two parts of the city of Hamilton characterized as being geographically diverse, possessing varying neighbourhood characteristics and qualities. These neighbourhoods are quite distinct from one another in terms of education, employment and income levels. They were chosen because their boundaries had been outlined and their characteristics analyzed in a previous investigation (Luginaah et al. 2001). Participants were asked questions regarding their feelings about their neighbourhood, their sense of belonging, whether or not their neighbourhood reflected their identity, their sense of rootedness and their feelings of neighbourhood attachment.

Thematic analysis was conducted on the focus group data. A thematic approach employed in the context of qualitative research enables broad themes within the data to be identified and organized (Burnard 1991). Recorded focus group data were transcribed, after which the transcripts were freely coded. This process of visual coding involved creating appropriate codes and recording them on hard copies of each transcript. Transcript coding was an iterative process; as new codes emerged they were applied to previous transcripts where relevant. An electronic list of codes and appropriate parent themes was maintained to ensure that subsequent transcripts were consistently coded. Once all of the transcripts were visually coded, the data were imported into NVivo, a qualitative analysis software program, and electronically coded. The software facilitated thematic analysis by organizing the data and allowing the codes to be sorted, classified and grouped. Each parent theme encompassed several related sub-themes.

The themes identified during the literature review and qualitative analysis were brought together to inform questionnaire item development using the facet design approach. Different members of the research team had conducted the literature review and the focus group data analysis, and each prepared a list of themes that emerged from their respective data. There was almost 100% consistency between the themes developed from the literature and those developed from the qualitative data. The themes were grouped according to seven sense of place components: rootedness, belonging, place identity, meaningfulness, place satisfaction, emotional attachment and physical environment relations.

Facet design is an approach used to structure research (Hackett and Foxall 1997), based on facet theory, the latter which was originally devised by Louis Guttman (Brown 1985). It is a process applied in fields as diverse as psychology (Breslin et al. 2006), education (Cohen 2004) and archaeology (Loy 2003) and is typically used for quite narrow, focused areas of study, such as: motives for moving house, or a pupil's worries during the transition from elementary to secondary school. Although facet design is an approach used to organize and structure research from beginning to end, we use the design component only, as our objective was to devise a survey instrument. In this case, the facet design process was used to facilitate item development by ensuring that each of a series of identified themes and behavioural and perceptual components deemed valuable were represented in the item set, through the use of a *mapping sentence*. Mapping sentences are used to create a comprehensive inventory of desired data. A mapping sentence contains three primary types of components or *facets*: a) the population – the population of persons being studied (here, neighbourhood residents); b) the content domain – which can contain many individual facets that describe the experience or perception of respondents, including a behaviour modality facet and themes or other components pertaining to the area of study (Elizur and Yaniv 2005), which in this case is sense of place (obviously, the fewer the facets, the more straightforward the item construction); and c) the common range – the range of answers from which a questionnaire respondent can choose (Canter 1985).

Stashevsky and Elizur (2002) identify the three behaviour modalities that often serve as core components of the facet design process when it is applied to human activity: affective, instrumental and cognitive; Elizur and Yaniv (2005) supplement these with two additional modalities, those being value and norm. Affective represents elements related to satisfaction, instrumental is related to actual behaviour, and the cognitive modality pertains to belief (Stashevsky and Elizur 2002; Elizur and Yaniv 2005). Value and norm are slightly more abstract and relate to importance and desire respectively (Elizur and Yaniv 2005). These five modalities were chosen for the current application of facet design.

Time, source and demand were other facets initially proposed in the project protocol that were not made 'formal' facets due to their comparative irrelevance. Although items were inevitably created that fell into these three categories, it was determined that it wasn't practical to *systematically* ensure that there was an item for each of these areas (for example, ensuring that there was an item for each theme and modality for time – capturing the past, present and future), as they would not be relevant for all facets.

Each facet is composed of individual components called elements, which in turn may be broken down into sub-elements. Figure 6.2 illustrates this relationship.

For example, the 'sense of place' content facet contains the sub-element 'rootedness', which has three sub-elements: 1) feeling 'at home'/freedom to do what is desired; 2) rooted in the neighbourhood (including staying in the neighbourhood,

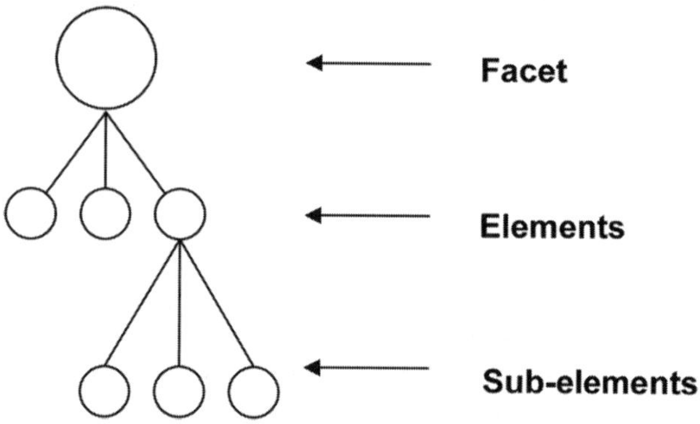

Figure 6.2 Facets, elements and sub-elements

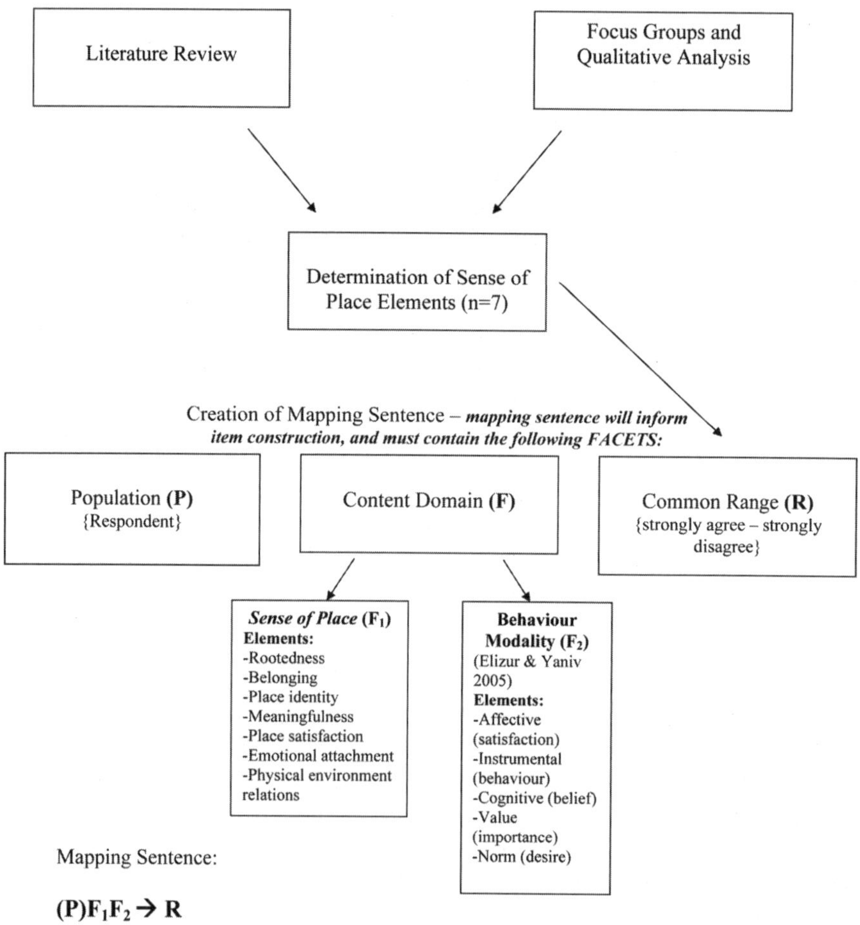

Figure 6.3 Item construction using facet design

feelings about moving, dependent but not rooted – e.g. for financial reasons, age, etc.); 3) familiarity/consistency/continuity.

There are several advantages and drawbacks to the facet design approach. As indicated, it facilitates the creation of a comprehensive list of items that covers all variables within the area of study (Hackett and Foxall 1997). The approach also enables items and material to be systematically accounted for and organized. There are also drawbacks to the process given that the determination of the components that will compose the mapping sentence is an abstract and subjective process (Elizur and Yaniv 2005). The variables included must be deemed relevant by the research team and while the mapping sentence itself is a formula, there is no set protocol for selecting its components.

Figure 6.3 illustrates the item construction process starting with the literature review and focus group data and ending with all of the variables contained in the mapping sentence.

As indicated earlier, the seven sense of place elements (F_1) (rootedness, belonging, place identity, meaningfulness, place satisfaction, emotional attachment, physical environment relations) and their sub-elements were developed from the literature review and focus group analysis. The behaviour modality elements (F_2) used were based on previous facet design work (Stashcvsky and Elizur 2002; Elizur and Vaniv 2005). Items using the value, norm and, to some extent, affective modalities, however, were in later parts of the project identified as not being very helpful to the overall goal of measuring sense of place. To illustrate, learning that a neighbourhood component is important to an individual (value), for example, or that he or she desires a particular component in a neighbourhood (norm), does not provide very much insight into whether or not these values or norms are actually achieved, and thereby do not have an impact on sense of place perceptions.

Below is the flow of the mapping sentence process (Figure 6.4), starting with the skeleton used and ending with a model questionnaire item.

The facet design approach utilizes Cartesian principles (Elizur and Yaniv 2005); in order to systematically create items for each theme, a matrix was created that was comprised of rows of sense of place elements and sub-elements and columns of the relevant behaviour modality elements. Members of the research team developed items cell by cell, discerning where each dyad would yield an appropriate item and where an item consisting of a particular combination of sense of place theme and behaviour modality would be/not be relevant. It was this stage in the process

Template statement: *The assessment of the participant {**P** pop$^{\underline{n}}$} of the {$\mathbf{F_2}$ behaviour modality} related to {$\mathbf{F_1}$ sense of place area} → {**R** range}*

Template sentence with specific elements: *The assessment of {**P** the neighbourhood resident} of the {$\mathbf{F_2}$ instrumental} related to {$\mathbf{F_2}$ place satisfaction} → {**R** strongly agree...strongly disagree}*

Example item: *I am involved in community activities I enjoy.*

Figure 6.4 Mapping sentence process

Table 6.1 Item construction table

Sense of Place Area	Affective (Satisfaction)	Instrumental (behaviour)	Cognitive (assessment belief)	Value (importance)	Norm (desire)
		The assessment of the participant{P} of the {behaviour modality} related to {sense of place area*} → {range}*			
ROOTEDNESS					
		I feel comfortable being myself in my neighbourhood	*I feel at home in my neighbourhood.* *I consider my house to be a home, not just a dwelling.*	*It is important for me to feel at home where I live.*	
- feeling at home/freedom to do what is desired. - rooted in the neighbourhood, feelings about moving, dependant but not rooted - e.g. for financial reasons, age etc.).	*If I had to move I would be disappointed.*	*If I had the chance to move, I would do so.* *I do not see myself moving out of the neighbourhood.*	*I feel strongly about remaining in the neighbourhood.* *There's no other place that I would rather live.*	*If I were to move, it would be important that I remain in the same neighbourhood.* *It is important for me to live somewhere where I can put down roots.*	*I would like to stay at home for as long as I am able to do so.*
- familiarity/ consistency/continuity (alternatively: changes)			*I know my neighbourhood like the back of my hand.* *My neighbourhood is constantly changing.*		

that made the approach taken for sense of place different from typical applications of facet design, where normally every cell would accommodate an item. In this particular application some boxes had one or more items while others held none. Additionally, instead of fitting each item to a set pattern or a sentence blueprint, each sentence was individually worded. Table 6.1 provides an example of this matrix chart, with items illustrated.

Typically, facet design is applied to areas of study that are much less broad and abstract than the concept of sense of place. In most of these cases, every behaviour modality is paired with *every* content facet component, often to create very uniform sentences. Depending on the subject matter, not all modalities are necessarily applicable, as demonstrated within the present application. Furthermore, the content facets and other components of the mapping sentence vary a great deal depending on the function of the facet design. Consider the following examples of mapping sentences used within the facet design process of other studies:

Components and Symbols of Ethnic Identity (Cohen 2004)

Ethnic group member **P** evaluates his/her ethnic identity in the {cognitive; affective; instrumental} **modality** in relation to the {specific; universal} **character** of the {biology; religion; history; culture; society; general} **domain** as referring to {Diaspora identity; homeland; host society; general} → the {low → high} **evaluation** of his/her ethnic identity.

Archaeological Tool Use (Loy 2003)

C {specific manufacturing technology} is done upon **A** {material} using **B** {tools} which exhibit **C** {structural/morphological attributes} and produce **D** {functional products} that exhibit **E** {structural/morphological attributes} which are some subset → of **P** {the population of manufacturing tools and processes}.

Because both of the above topics are limited in scope, it is possible to see how a very specific and uniform set of items could be constructed, and how these would vary greatly from the more flexible sense of place mapping sentence (and resultant items) developed for this project.

Transition from Item List to Measurement Tool

The initial item list was based on a number of broad elements and sub-elements and consequently contained many times the number of items that were required for the end product. This was anticipated and multiple strategies were employed to pare down the number of items. The first strategy was to engage the full research team in discussion of each item, where group consensus was used to reduce the pool of items from 194 to 112.

Next, a cross-disciplinary five-member expert panel was assembled to provide feedback about item relevancy and content coverage. This panel was made up of individuals from various research backgrounds who had extensive experience working in sense of place or related areas. The disciplinary backgrounds of the expert panel members included: sociology and education, environmental psychology, forestry, rural sociology and natural landscapes sciences. Those who agreed to participate were given an evaluation form containing all of the draft items. Participants were asked to rate the relevancy of the items to sense of place on a five-point scale, with 1 being not at all relevant and 5 being extremely relevant. Subsequently, each expert panel member participated in a one hour telephone conversation with a member of the research team, to discuss their ratings, identify areas of redundancy and describe any areas or items that appeared to have been omitted. Once input had been received from all of the members of the expert panel, the ratings for each item in the pool were assessed to determine which should remain. Those items that received an average rating of 4 or above, as well as those items where the majority of items received ratings of 4 or above, were retained. This phase of the project enabled the item list to be further pared down from 112 to 57.

The next step was to explore how individuals interpret and speak about sense of place (i.e. check the language used by lay persons) through cognitive interviewing. For efficiency, the remaining items were divided into three pools with three pseudo-surveys created. Six one-on-one interviews were conducted with community members from the neighbourhoods where the original focus group participants lived. The six participants completed one of the pseudo-surveys composed of a different third of the item pool (two participants were interviewed about questions on all three pseudo-surveys). All participants were asked a series of questions after each item was presented/asked. These questions were chosen in order to expose items that were ambiguous, incomprehensible, unclear or otherwise not applicable to community members. Before proceeding to the next phase of the project, the members of the full research team came together to discuss the results and recommendations of the expert panel members and the cognitive interviews. The team discussed and revised the final draft list of items, deciding on 48 items for the final survey tool.

Lastly, a selection of health items were added in order to enable an analysis of the relationship between sense of place and health; this included the excellent-very good-good-fair-poor (EVGFP) scale to measure self-reported health (Ware and Sherbourne 1992). This final questionnaire survey was then submitted to an external social research institute, contracted to collect primary survey data via mail for psychometric testing.

Conclusion

As a result of the application of facet design to this project, the item construction process was systematic, efficient and most importantly, comprehensive. Although the initial item list created inevitably had to be pared down to a smaller number of items that was more practical for a questionnaire survey, we can be sure that we had encompassed all of the essential *sense of place* components before we proceeded to the next stage of the project. The flexibility in the mapping sentence process enabled us to fit all of the elements and sub-elements identified as important, as confirmed via the focus group results and literature review.

There are some limitations to the use of facet design within a broad construct such as sense of place. Within this application, and despite the removal of irrelevant modalities, there was a great deal of thematic repetition within the initial item pool. Further, because of the broad nature of sense of place and our desire to encompass all elements of it, the facet design process yielded a very large number of items. Both of these results required in-depth discussion among members of the research team in order to determine ultimate applicability of each item within a reasonably sized final pool.

Researchers involved in projects pertaining to broad subject areas like sense of place should not be put off by the constricted appearance of the facet design process. As demonstrated in this chapter, the process can be modified and successfully applied to wide-ranging topics of study which are characteristically broad. It should be noted that, like other theories and approaches to research design, the facet design process is a *tool*. When applied appropriately it can facilitate project progress. The facet design process will not provide the user with results without: considerable contemplation by the research team, further input from individuals such as those within the expert panel, cognitive interviews with potential respondents, and ongoing psychometric testing.

Within this tool development project, the facet design process has transformed an assembly of qualitative data consisting of community focus group data and academic literature into a comprehensive set of items reflective of key sense of place elements and themes. Cognitive interviews and specialist input in the form of expert panel discussions and research team contributions yielded the items that would be most appropriate for the measurement tool. Psychometric testing of the large community pilot test will play the last key role in achieving the goal of developing a tool that can validly and reliably measure sense of place.

References

Altschuler, A., Somkin, C.P. and Adler, N.E. (2004), Local services and amenities, neighborhood social capital, and health, *Social Science and Medicine* 59: 1219–1229.

Breslin, F.C., Hepburn, C.G., Ibrahim, S. and Cole, D. (2006), Understanding stability and change in psychological distress and sense of coherence: A four-year prospective study, *Journal of Applied Social Psychology* 36(1): 1–21.

Brown, J. (1985), An Introduction to the Uses of Facet Theory. In Canter, D. (ed.) *Facet Theory: Approaches to Social Research* (New York: Springer Verlag New York Inc.).

Burnard, P. (1991), A method of analysing interview transcripts in qualitative research, *Nurse Education Today* 11: 461–466.

Canter, D. (1985), How to be a Facet Researcher. In Canter D (ed.) *Facet Theory: Approaches to Social Research* (New York: Springer Verlag New York Inc.).

Clark, J. and Stein, T.V. (2003), Incorporating the natural landscape within an assessment of community attachment, *Forest Science* 49(6): 868–876.

Cohen, E.H. (2004), Components and Symbols of Ethnic Identity: A Case Study in Informal Education and Identify Formation in Diaspora, *Applied Psychology* 53(1): 87–112.

Cresswell, T. (1996), *In Place/Out of Place: Geography, Ideology and Transgression* (Minneapolis, MN: University of Minnesota Press).

Dunn, J. (2002), Housing and inequalities in health: a study of socioeconomic dimensions of housing and self reported health from a survey of Vancouver residents, *Journal of Epidemiology and Community Health* 56: 671–681.

Elizur, D. and Yaniv, E. (2005), Advantages and limitations of Facet Theory compared to other approaches. Accessed online: http://www.europhd.psi.uniroma1.it/html/_onda02/07/PDF/Elizur,D.(2005).ppt.

Eyles, J. (1985), *Senses of Place* (Warrington: Silverbrook Press).

Fletcher, C.M. (2006), Environmental Sensitivity: Equivocal Illness in the Context of Place, *Transcultural Psychiatry* 43(1): 86–105.

Frumkin, H. (2003), Healthy places: exploring the evidence, *American Journal of Public Health* 93(9): 1451–1456.

Gesler, W. (1992), Therapeutic landscapes: Medical issues in light of the New Cultural Geography, *Social Science and Medicine*, 34:735–746.

Hackett, P. and Foxall, G.R. (1997), Consumers' evaluations of an international airport: a facet theoretical approach, *Review of Retail, Distribution and Consumer Research* 7: 341–351. ISSN 0959-3969.

Hartig, T. and Staats, H. (2003), Guest editors' introduction: Restorative environments. *Journal of Environmental Psychology* 23: 103–107.

Hay, R. (1998), A rooted sense of place in cross-cultural perspective, *Canadian Geographer* 42(3): 245–266.

Healey, P. (1998), Institutionalist theory, social exclusion and governance. In: A. Madanipour, G. Cars and J. Allen (eds), *Social Exclusion in European Cities*, pp. 53–74 (London: Jessica Kingsley Publishers).

Herzog, T.R., Maguire, C.P. and Nebel, M.B. (2003), Assessing the restorative components of environments, *Journal of Environmental Psychology 23*: 159–170.

Kearns, A., Hiscock, R., Ellaway, A., and Macintyre, S. (2000), 'Beyond four walls'. The psycho-social benefits of home: evidence from west central Scotland. *Housing Studies*, 15:387–410.

Kianicka, S., Buchecker, M., Hunziker, M, and Muller-Boker, U. (2006), Locals' and tourists' sense of place: A case study of a Swiss Alpine village, *Mountain Research and Development* 26(1): 55–63.

Korpela, K. and Hartig, T. (1996), Restorative qualities of favorite places, *Journal of Environmental Psychology* 16: 221–233.

Loy, T.H. (2003), Chapter 2: On the Explanation of Culture Change: Theory, Methodology and Methods. Course notes: ARCA3000, School of Social Science, University of Queensland.

Luginaah, I., Jerrett, M., Elliott, S., Eyles, J., Parizeau, K., Birch, S., Abernathy, T., Veenstra, G., Hutchinson, B. and Giovis, C. (2001), Health profiles of Hamilton: Spatial characterisation of neighbourhoods for health investigations, *GeoJournal* 53: 135–147.

Manzo, L. (2003), Beyond house and haven: Toward a revisioning of emotional relationships with places, *Journal of Environmental Psychology* 23: 47–61.

Macintyre, S., Maciver, S. and Sooman, A. (1993), Area, class and health: Should we be focusing on places or people? *Journal of Social Policy*, 22: 213–234.

McDowell, L. (1997), *Undoing Place? A Geographical Reader* (London: Arnold).

Meegan, R. and Mitchell, A. (2001), 'It's not community round here, it's neighbourhood': neighbourhood change and cohesion in urban regeneration policies, *Urban Studies*, 38(12): 2167–2194.

Mitchell, R., Gleave, S., Bartley, M., Wiggins, D. and Joshi, H. (2000), Do attitude and area influence health? A multilevel approach to health inequalities, *Health and Place* 6: 67–79.

Oakes, T. (1997), Place and the paradox of modernity, *Annals of the Association of American Geographers*, 87: 509–531.

Pretty, G.H., Chipuer, H.M. and Bramston, P. (2003), Sense of place amongst adolescents and adults in two rural Australian towns: The discriminating features of place attachment, sense of community and place dependence in relation to place identity, *Journal of Environmental Psychology*, 23: 273–87.

Relph, E. (1976), *Place and Placelessness* (London: Pion).

Rioux L. (2005), The well-being of aging people living in their own homes, *Journal of Environmental Psychology* 25: 231–234.

Shamai, S. (1991), Sense of Place: an Empirical Measurement, *Geoforum* 22(2): 347–358.

Stashevsky, S. and Elizur, D. (2002), Quality management of organizational referents: A structure analysis, *International Journal of Manpower* 23 (4): 362–375.

Stedman, R.C. (2003a), Sense of place and forest science: Toward a program of quantitative research, *Forest Science*, 49(6): 822–829.

Stedman, R.C. (2003b), Is it really just a social construction? The contribution of the physical environment to sense of place, *Society and Natural Resources*, 16: 671–685.

Theodori, G.L. (2001), Examining the effects of community satisfaction and attachment on individual well-being, *Rural Sociology*, 66(4): 618–628.

Tuan, Y. (1974), *Topophilia: A Study of Environmental Perception, Attitudes and Values* (New Jersey: Prentice-Hall).

Ulrich, R.S. (1984), View through a window may influence recover from surgery, *Science* 224: 420–421.

Ware, J.E. and Sherbourne, C.D. (1992), The MOS 36-Item Short-Form Health Survey (SF-36). I. Conceptual Framework and Item Selection, *Medical Care* 30(6): 473–481.

Warin, M., Baum, F., Kalucy, E., Murray, C. and Veale, B. (2000), The power of place: Space and time in women's and community health centres in South Australia, *Social Science and Medicine*, 50:1863–1875.

Wells, N.M. (2005), Our housing, our selves: A longitudinal investigation of low-income women's participatory housing experiences, *Journal of Environmental Psychology* 25: 189–206.

Williams, A. (1998), Therapeutic landscapes in holistic medicine, *Social Science and Medicine* 46(9): 1193–203.

—— (2002), Changing geographies of care: employing the concept of therapeutic landscapes as a framework in examining home space, *Social Science and Medicine* 55(1): 141–154.

Wilson, K., Elliott, S., Eyles, J. and Keller-Olaman, S. (2007), Factors affecting change over time in self-reported health, *Canadian Journal of Public Health* 98(2): 154–8.

Yung, L., Freimund, W.A. and Belsky, J.M. (2003), The politics of place: understanding meaning, common ground, and political difference on the Rocky Mountain Front, *Forest Science*, 49(6): 855–866.

Chapter 7

The Experience of Displacement on Sense of Place and Well-being

Lynne C. Manzo

From the cover, she stares out at me – a forlorn looking African-American woman holding a baby. She is slumped in a chair in a room with deep cracks in the wall, peeling plaster and a leaking radiator. A tattered calendar hangs lopsidedly from the wall with the date August 8, 1992 clearly marked. Behind her is a window through which I can see a new two-story building with shiny windows and a front lawn. The number over the doorway reads "8-2000." On the walkway in front of the building stands the very same woman with her hair neatly combed back. She appears dressed for work. In one hand she carries a briefcase; in the other, she holds the hand of a young child clutching books, ready for school. I am looking at the cover of the Final Report of the U.S. National Commission on Severely Distressed Public Housing. It is a dramatic sketch that speaks volumes about our beliefs on poverty, public housing and place.

Released in August 1992, this report was the result of a two-year investigation into the condition of public housing in the United States, and it marked the beginning of a dramatic shift in public housing policy that has altered the landscape of affordable housing and changed the lives of tens of thousands of public housing residents nationwide. As a result of the report, the HOPE VI program was created. Meant to eradicate what the Commission considered "severely distressed" public housing by the year 2000, HOPE VI (Housing Opportunities for People Everywhere) aims to disperse pockets of poverty and transform public housing sites into new mixed-income developments.

It is the vision – and assumptions – of this program that we see depicted on the cover of the Commission's Final Report. Through the window in the sketch we are meant to see a glimmer of hope, eight years into the future. It is August 2000; the HOPE VI redevelopment of our protagonist's housing is complete, and she can look forward to a bright future with new housing, a job and a child headed to school. As an environmental psychologist and a housing researcher interested in social justice, I wondered: What would it take to realize this vision? Would the physical transformation of public housing lead to life improvements for residents? What is gained and what is lost in such a transformation?

The realization of the Commission's vision to eradicate severely distressed public housing necessitates the displacement of residents as public housing sites are demolished and rebuilt as mixed income communities. The magnitude of the displacement is considerable. To date, 609 HOPE VI grants have been awarded to

Public Housing Authorities (PHAs) in 193 cities (Urban Institute 2007). As of June 2006, a total of 78,100 units were demolished, with another 10,400 units slated for redevelopment (The Urban Institute, 2007). While HOPE VI will construct 103,600 replacement units, just 57,100 will be deeply subsidized public housing units; the rest will receive shallower subsidies or the units will go to market-rate tenants or homebuyers (Urban Institute, 2007). This means fewer original residents will be able to return to redeveloped sites, causing permanent displacement. With low rates of return, some have argued that HOPE VI has not met its original vision of residents returning to revitalized units and that for most original residents the major impact of HOPE VI is relocation (Popkin 2007).

This reality of displacement nation-wide as well as *who* is being displaced raises important questions, as public housing residents are among the poorest households in the nation, and are the most vulnerable to housing insecurity, health problems and other challenges to their overall well being. In terms of poverty, the vast majority of public housing applicants nation-wide (89 per cent) earn less than 30 per cent of the Area Median Income (AMI), making them what is considered "extremely low income."[1] In addition, many public housing sites include families that are considered "hard to house" i.e. families with multiple complex problems including physical disability, substance abuse, large numbers of children and a weak labor market history (Cunningham et al., 2005; Popkin 2007). Moreover, poor health is a considerable challenge for public housing residents. Adults in studies of HOPE VI outcomes experience health problems more often than other demographically similar groups (Manjarrez et al., 2007).

Given this context, we must examine resident's lived experience of place to determine whether they consider their housing and communities distressed, and to consider the social, emotional and health costs of dismantling existing communities. Decades of research in social epidemiology show clear connections between housing and health, and between social connectedness and health (Berkman and Glass 2000; Berkman and Kawachi 2000; Dunn 2002; Shaw 2004). Likewise, we need to understand the full consequences of uprooting HOPE VI residents from their housing and their community of neighbors.

This chapter first examines the literature on the lived experience of place and its connections to health then draws on data from a case study of one HOPE VI community before it was redeveloped in order to consider what is at stake from displacement. The case study research employed a mixed methods approach including a needs assessment survey of all household heads (n=512), focus groups held in the three main languages spoken on site (Vietnamese, Cambodian and English) and in-depth

1 AMI is an annual calculation provided by the US government for each community across the nation. It divides the household income distribution into two equal parts where half of the cases fall below the median household income and half fall above the median (Greater New Orleans Community Data Center, accessed online July 26, 2007). Typically it is adjusted for family size. According to the U.S. Dept of Housing and Urban Development (HUD) any household earning 80 per cent or less of the AMI is considered low income and is eligible for public housing. However, in practice, the vast majority of public housing residents earn no more than 30 per cent of the AMI.

qualitative interviews with residents in English (n=47).[2] This study revealed that prior to redevelopment this housing site was a socially well-functioning community where residents laid down roots and formed place attachments and bonds of mutual support with neighbors (Manzo, Kleit and Couch 2008). In this chapter I consider the implications that such findings have for the health and well-being of HOPE VI residents and suggest areas for further examination.

The Lived Experience of Place

Residents' lived experience of place is an essential ingredient in understanding whether and how public housing sites might be distressed. Indeed, the focus on outcomes in HOPE VI research is largely to determine whether physical environmental improvements can improve life circumstances for public housing residents more broadly and contribute to greater well-being (Popkin 2007). The inclusion of the lived experience of place in this research will enable policymakers to make more informed decisions about whether and how to redevelop housing for the poor. Insights from the resident perspective enable us to view "the projects" not as abstract politicized places of distress, but as meaningful places where everyday life unfolds, that is, as home (Manzo, Kleit and Couch 2008). Let us first review a few critical concepts at the core of the lived experience of place: sense of place and place attachment.

Sense of Place

To begin, place is a fundamental ontological structure; it is where life is lived both instrumentally and symbolically (Butz and Eyles 1997). Early work on sense of place differentiated place from space as centers of felt values (Tuan 1976, 1977), and indeed of human existence (Relph 1985). One early definition of sense of place describes it as "an experiential process created by the setting, combined with what a person brings to it" (Steele 1981, p. 9). This definition underscores the transactional and dynamic nature of our relationships to place. Somewhat later, Hummon (1992) describes sense of place as involving an interpretive perception of the environment and an emotional reaction to that environment. Since this earlier work, the sense of place construct has undergone additional extensions, revisions, and re-evaluations (see Farnum, Troy and Kruger 2005 and Chapter 1 in this volume for an overview). For example, it has been expanded to include: "the social and historical processes by which place meanings are constructed, negotiated and politically contested" (Williams and Stewart 1998: 20). Similarly, Butz and Eyles (1997) extend the understanding of sense of place in humanistic geography by including concepts from political geography and interpretive anthropology. In doing so, they identify

2 This research was conducted in collaboration with Rachel Garshick Kleit from the Evans School of Public Affairs, University of Washington, Seattle, U.S. The author of this chapter was the principle investigator on the project. To protect the privacy of both residents and Housing Authority staff, I do not reference the site nor the agency involved nor the original evaluation report to the HA.

three core components of sense of place – social, ideological and ecological – that reflect its more nuanced, complex nature as constituted through socio-political and symbolic processes as well as material circumstances.

There is some question about how sense of place relates to other concepts such as place attachment and place identity. Elsewhere I have expressed my concern for greater conceptual clarity of sense of place as it is used in current empirical research (Manzo 2003, 2005). Like others, I contend that sense of place is broader than place attachment (Hummon 1992; Butz and Eyles 1997; Hay 1998) and I use sense of place as the more general term, referring to both affective and cognitive components of place. Indeed, the most recent conceptualizations and measurements of sense of place include rootedness, belonging, place identity, meaningfulness, place satisfaction, emotional attachment and physical environmental dimensions as component parts (Williams, Eyles and Heidebrecht 2008).

Place Attachment

Place attachment can be considered a dimension of sense of place. Generally, it is an emotional bond between people and places (Altman and Low, 1992) that operates at both the individual and community levels (Hummon 1992). On the individual level, place attachments reflect our embeddedness in our everyday surroundings, behaviorally, emotionally and cognitively (Brown and Perkins 1992). On the community level, attachments reflect a sense of bondedness to the social community and a sense of rootedness in the physical community (Riger and Lavrakas 1981). It is noteworthy that local social involvements are the most consistent and significant source of sentimental ties to local places (Hummon 1992) and that these attachments, in turn, lead to greater social cohesion and social control (Brown, Perkins and Brown, 2003). Such attachments are influenced by length of residency, and are higher among elderly and families with children (Hummon 1992).

Studies of place attachment on the community level reveal that neighborhoods are important socio-spatial contexts in which a sense of place develops. Eyles (1985) identifies three elements of community that enriches our notion of place: the physical environment itself, people and their institutions, and a sense of belonging. This three-pronged approach can help "provide insights into the importance and role of place in social and material life" (p. 63). Indeed place is not a mere container of experience; it is part of people's lived experience. Similarly, people do not just live in houses; they dwell in and experience neighborhoods (Shaw 2004, p. 412). This shared emplaced reality leads to the symbolic component of people-place relationships – the expression of collective sentiment (Butz and Eyles 1997).

Most famously we know of the place attachments illustrated in the classic studies by Fried (1963) and Gans (1962) on inner city working class neighborhoods that show how these neighborhoods, in contrast to outside perceptions of them as distressed, served as a locus of meaning and value for residents. Recent research shows that place attachments have potential strength in declining neighborhoods that is often overlooked but that can be utilized to mobilize revitalization efforts (Brown, Perkins and Brown 2003). In their study of over 600 households in a neighborhood in decline, Brown, Perkins and Brown (2003) found that residential attachments promote and provide stability, familiarity, and security – critical ingredients in psychological well being. They, along with Fried (2000), argue that place attachments may be especially strong in lower income and/or ethnic neighborhoods, given their relative isolation from the larger society and the wide range of supports they provide residents (Brown, Perkins and Brown 2003, p. 261).

While poor neighborhoods may be considered places of social isolation because of their economic homogeneity, and those populated largely by immigrants and refugees may experience an additional layer of isolation from the larger, mainstream society, there is evidence that this homogeneity leads to greater degrees of place attachment and interdependence among residents (Fried 2000). The question is whether place attachments in this context are helpful or limiting in the long run. However, attachments are responsive to human aspirations and don't necessarily hinder them. They serve critical instrumental and emotional functions. Moreover, attachments are dynamic and can adjust to changes if given the right opportunities e.g. adequate transition time, rituals to acknowledge and honor place meanings, and allowing choice in relocation that help mitigate a feeling of helplessness (Fried, 2000). This is where we need to be particularly careful about what kind of outcomes we attend to when thinking about the impacts of attachments and displacement. For example, relocation might help meet more individualistic goals in some cases such as improved status and economic well being (Brown and Perkins 1992) – but it may disrupt more socially-oriented goals and mutual support systems.

Literature from social epidemiology would not necessarily consider public housing sites with a good deal of mutual support among neighbors to be socially isolated, as here isolation is determined by whether there are opportunities for social engagement, and whether people partake in these opportunities regardless of the demographic similarities or differences from larger society (see especially Berkman and Glass 2000). It follows then that neighborhoods in which residents engage with one another – for example, HOPE VI sites where resident experience place attachments and mutual support – can be considered socially healthy and well functioning. A closer look at the social epidemiology literature can shed valuable light on both the health implications of residents' place experiences and of relocation.

Sense of Place, Health, and Well-Being

Social epidemiology has shown us that health is influenced by factors not only on the individual level but the community level. For example, Beekman and Glass (2000) argue that the degree to which an individual is interconnected and embedded in a community is vital to both the individual's health and well-being, and that of the entire community. From this perspective housing is not only a resource and material good (the condition of which has an important, direct influence on health), it is also the location for social networks. These ties give meaning to our lives – and to place – by enabling us to participate more fully in it (Beekman and Glass 2000). Social networks also influence health in important ways – through the provision of social support, through social influence and through opportunities for social engagement (Beekman and Glass 2000).

In particular, social support exchanges can be powerful facilitators of health and well being. Beekman and Glass (2000) identify four main types of social support: emotional, instrumental, appraisal and informational. The latter three influence health by improving access to resources and material goods. The emotional component provides a sense of security, belonging, self esteem and a greater resilience to life challenges. Support exchanges based on a shared history are particularly powerful. This is certainly the case with HOPE VI residents who share the common bond of living in public housing and are all trying to get by on extremely limited resources.

In terms of the connection between material environment and health, it is certainly true that many public housing sites have fallen into considerable physical disrepair and these poor physical conditions have health implications. Mould and mildew in the home, a widespread problem in much public housing, are connected to higher rates of asthma, while exposure to lead paint (used in the construction of many public housing sites decades ago) has considerable physical and psychological health effects. But direct effects of the material condition of housing are no longer the only way in which housing and health are related (Shaw 2004: 406). The symbolic and emotional meaning of one's housing is also critical.

Place Attachment and Well-being

Sense of place and residential places attachments are connected to a sense of well being in important ways (Butz and Eyles 1997; Harris, Werner, Brown and Ingebritsen 1995). Well-being is an elastic and multi-dimensional concept that refers to a variety of phenomena; most broadly it has been considered to be whatever is assessed in an evaluation of a person's life situation (McGillvray 2007). It has been equated with life satisfaction, basic need fulfillment, and happiness. One of the most influential conceptualizations of well being is known as the capabilities approach which connects well-being to a person's capability to achieve valuable functionings (see esp Sen, 1993).

Related to this approach to well-being is Wallenius' (1999) notion of "personal projects" – our goals and the activities in which we engage to achieve those goals. These range from more mundane activities (going to the store) to loftier endeavors (getting a new job). While this work is not part of the well-being literature, it is quite commensurate with the capabilities approach to well-being because it concerns itself with what people do to meet their goals, i.e. it focuses on functioning and how the environment can support our ability to meet our needs. The idea of personal projects is very helpful when considering the value and role of a public housing community for its residents. Like anybody else, most public housing residents are trying to meet their basic needs and goals as best they can. Place attachments and social support among neighbors can greatly facilitate those goals. Indeed, Wallenius' view of personal projects includes a spatial dimension in that it has to do with the perceived supportiveness *of the environment* in meeting our goals. That is, places themselves gain meaning not just as the backdrop in which our personal projects are carried out, but as "active agents" in the project themselves (Sack 2001, p. 232; Easthope 2004). Those places we feel are supportive of our personal projects are among those to which we become attached.

It is also important to consider the perceived supportiveness of the environment on the community level. In addition to personal projects there are shared efforts and goals among neighbors to carry out their daily lives as best they can. This is what Fullilove (2004) and Levy (2005) call "the common life." Among public housing residents, who have limited resources along with complicated personal and housing histories, the common life provided in their community can be particularly supportive and stabilizing.

Not only is it important to consider the instrumental functions of these social ties and place attachments but to consider the symbolic and psychological meaning of housing for well being. Housing is one of the key social determinants of health, not simply in terms of provision of shelter or the material condition of that shelter, but in terms of the meaning of housing and its purpose in providing "ontological security" or a sense of security, control and mastery (Shaw 2004). While the direct effects of material conditions are critical (e.g. mold and respiratory health), these are not the only way that housing and health are related. What is also important is the meaning of home *as a place of constancy*, [emphasis mine] and a feeling of centeredness and belonging (Shaw 2004). One large study in Vancouver, British Columbia reveals that control in and over one's place of residence is associated with well-being and that those worried about being forced to move were significantly more likely to report poorer physical and mental health (Dunn 2002).

Disruption, Displacement and Implications for Well-Being

As we consider attachment and sense of place, so too must we address displacement. Certainly the experience of displacement, especially among the poor, is not new. It occurs worldwide and continues as a result of urbanization and gentrification processes (Manzo Kleit and Couch 2008). While the HOPE VI program is distinct from these other displacement scenarios, and Housing Authorities like the one redeveloping the site of this case study are actually prioritizing the return of original residents, we still do not know the full implications and outcomes of relocation. Decades of research have demonstrated displacement's significant impacts on well-being and health. Forced displacements "are among the most serious forms of externally-imposed psychosocial disruptions and discontinuities" (Fried 2000, p. 194). Is the same true for HOPE VI?

Large-scale social upheavals and transformations profoundly disrupt patterns of social organization. (Berkman and Glass 2000). "Geographic relocations related to housing policy are among the environmental challenges that tear at the fabric of social networks and in turn have deleterious effects on health" (ibid, p. 153). When environments change rapidly and when relocations are involuntary, place attachments are even more disrupted (Brown and Perkins 1992). Involuntary relocations such as those brought about by economic development initiatives are sudden and the change threatens to overwhelm stability (Brown and Perkins 1992). Such disruptions disturb a sense of continuity and cause feelings of loss and alienation (Hummon 1992).

As a psychiatrist, Fullilove (1996, 2004) has studied the psychological impacts of forced relocation through urban renewal programs, demonstrating that the loss of the "anchoring community" can cause disorientation, confusion, depression and physical health problems related to stress. This phenomenon, which Fullilove calls "root shock," was not only experienced by residents of neighborhoods demolished by urban renewal, but also by residents of a HOPE VI redevelopment in Pennsylvania which she studied. Root shock poses a distinct threat to well-being; it is "the traumatic stress reaction to the destruction of all or part of one's emotional ecosystem" (Fullilove 2004, p. 10). At the level of the individual it is a profound upheaval that destroys one's working model of the world upon which we depend and act in our everyday lives. When our working model of the world is damaged and destroyed, it threatens our ability to complete our personal projects – that is, to get things done and live comfortably well – because our working model is built on our perceptions of the supportiveness of places in our everyday lives to meet our goals and emotional needs. If those places are destroyed or our access to them is cut off, it threatens our well being.

Newman and Wyly (2006) argue that "a new generation of quantitative research has provided new evidence of the limited (and sometimes counter-intuitive) extent of displacement, supporting broader theoretical and political arguments favoring mixed-income redevelopment and other forms of gentrification" (p. 23). This evidence is being used to dismiss concerns about a wide range of market-oriented urban policies of privatization, home-ownership, 'social mix' and dispersal strategies designed to break up the concentrated poverty that has been taken as the shorthand explanation for all that ails the disinvested inner city (Newman and Wyly 2006). One such strategy is the HOPE VI program. In order to truly weigh the costs and benefits of the program, we must consider what kind of community was in place before demolition and how residents felt about this community to ascertain what may be lost and the impact this loss may have. A closer look at residents' lived experience of place in one HOPE VI community is offered in this chapter to get a fuller understanding of the role it played in residents' lives.

The Resident Perspective – Highlights of Case Study Findings

It is first helpful to get a sense of the physical community of this case study. In contrast to public housing sites in the eastern portion of the US which were built as high density hi-rises, the housing on this site was only modestly dense, one-storey duplexes – very much like the other public housing developments in the region. The housing on this site was built around the time of World War II as temporary housing for defense workers and their families, including women whose husbands were soldiers overseas. The units were barracks-style duplexes scattered on ill-defined open space. Given the wet climate in the region, mould and mildew and stormwater runoff evolved over time as problems, and this was part of the basis for seeking a redevelopment grant from the government. The redevelopment is still ongoing; at the time of this writing the construction is near completion.

So who were the people who lived on site at the time the redevelopment began some 60 years later? No longer "Rosie the Riveter" families, this recent population was composed of poor and largely immigrant families. The typical household contained three people. Over half were family households containing at least one child and an adult. Household heads ranged in age from 19 to 94. Sixty percent of all household heads were women and 36 per cent were seniors and/or disabled. All of the households were extremely low income, earning less than 30 per cent of the AMI. The average household income was $12,000, but half of the households earned less than $10,000 annually. The average household had lived in housing run by this Housing Authority for about 5 years, but the range is considerable. Some had moved onto the site shortly before the HOPE VI grant was awarded but several household heads had lived on site for more than 35 years. This same distribution appeared in the smaller interview sample (n=47). Here, 52 per cent had lived on site for 2–10 years and fully 39 per cent had lived there for 11–20 years.

A Decent Home, a Stabilizing Force

Given the poverty levels and the length of residency, it is not surprising that residents found their housing to be a stabilizing force. In contrast to the stigmatized views of public housing as being "housing of the last resort," all residents had positive things to say about the housing development and the role it has played in their lives (Manzo, Kleit and Couch 2008). Instead of being housing in which residents feel trapped, public housing residents themselves tend to view their communities as permanent places of residence (Vale 1997). As one resident explained, "We were safe and warm and we were stable. And this is where I want to be." When they were asked to compare their community with the surrounding neighborhood, over half of all residents (60 per cent of 507) thought it was a better place to live, a quarter (26 per cent) thought it was about the same and only 1 per cent thought it was worse. The prevalence of positive sentiments was also evident in the interviews. The majority of interviewees (64 per cent of 47) thought that the housing development was "a good place to live" or said they had a "good life" there. (Manzo ct al. 2008). It was a place that helped families to meet their basic needs and carry out their personal projects in three areas: residential security, financial stability, and a good life for their children. Many residents commented that it provided them with a decent, stable home and a way to keep their families living together:

> It was a place for us all to be together, and the kids to grow up and go to school and then go on with their own lives. And I, you know, if we'd lived somewhere else and couldn't afford it, I'd probably couldn't have kept my children, you know?

The housing development contributed to family well-being and allowed parents to provide their children "a decent home to grow up in." Residents commented that it was "a family place" a place that was "good for kids."

> If anybody would ask me on the street what kind of place it is, I would just tell them honestly, it's a united place where your family could be together, work out whatever problems you have. It's the kind of place that helps.

For many, their housing was an alternative to rather grim circumstances – homelessness or unlivable housing conditions.

> Well to me it's meant a lot – It's meant having a roof over my head. And a place for my son to come. It means a lot because for a minute I didn't think I was going to have somewhere to go after I had my heart attack. You know [I was] homeless and everything. So it really meant something to me to have a roof over my head. [This place] helped me out with that. I appreciate that.

For these residents in particular, the community had been a stabilizing force in their lives. As another resident commented:

> My experience [before] moving here, for, like, three or four years, I, I, it was so stressful, just putting kids through school. And I barely … was able to buy things for myself, you know, and so this really has been a blessing. So I'm a person, I'm grateful.

Clearly, this housing meant much more than simply shelter; it enabled residents to move toward their life goals (Manzo, Kleit and Couch 2008), and in that regard, contributed to well being and effective functioning.

The rich qualitative data describing residents' feelings about and experiences in their home and community were supported by rating scales. When asked to rate their attachment on a scale of 1 to 10, with 1 being not attached and 10 being strongly attached, scores averaged around 8, reflecting significant attachments to both their individual housing unit and the community. Residents also responded to a series of statements meant to indicate attachment on the block level. These statements, used in other research on residential attachment, included: (1) I think my block is a good place for me to live; (2) I feel at home on this block; (3) It is very important to me to live on this particular block; (4) I can recognize most of the people who live on this block; and (5) People on this block do not share the same values. Most people

Table 7.1 Attachment to the original block (n=502)

	Mean	Median	75% Percentile
I feel at home on this block	9	9	10
I think my block is a good place to live	9	9	10
It is very important to me to live on this particular block	8	9	10
I can recognize most of the people who live on my block	7	7	9
People on this block do not share the same values	5	5	8

Respondents were asked whether, on a scale of 1 to 10, they agreed or disagreed with the statements about their blocks. 1 was strongly disagree and 10 was strongly agree. Most respondents agreed strongly with all these statements. Less agreement was evident regarding sharing values where half the respondents were at least neutral. The Cronbach's alpha of scale reliability increased from .56 to .67 when the item regarding values was omitted.

agreed strongly with the first four (positively-valenced) statements and disagreed with the last (negatively-valenced) statement, indicating that respondents tend to believe that their neighbors do share the same values despite any cultural differences among immigrants.

Satisfaction or Satisfice?

Overall, residents rated their satisfaction with their housing unit, their immediate block, and the neighborhood quite highly. On average, residents rated their satisfaction with each at an 8 on a scale of 1 to 10, with 1 being not at all satisfied and 10 being completely satisfied.

Certainly we must be cautious about how to interpret these ratings. Research on residential satisfaction suggests that an analysis of ratings should include the rationale behind a person's score (Vale 1997). A closer look at interviewees' explanations sheds light on why they rated their housing the way they did (i.e. what they focus on and why), and how residents' expectations impact satisfaction. For example, of the interviewees who rated satisfaction with their unit (n = 42), 26 per cent rated it a "10" which was the most frequently occurring score. Scores ranged from 2 to 10, and the average score was a 7.7. Their reasons show that at least for some, their expectations are modest. As one elderly resident explained:

> This house … is fine, I'm satisfied. Because, I'm old, I've lived long and it's enough. When my son goes to school, gets out and buys his house, that's another story, but I'm old, living like this is enough.

We can see a similar dynamic with block satisfaction. When one resident asked why he rated his block a 10, he explained, "It's quiet, clean and no violence." Another described how their neighbor in their previous community dealt drugs and trashed the area outside. In comparison, he thought his current block was "very peaceful." Fried (2000) argues that people with limited resources and housing options may only hope for their residence to "satisfice" – that is, if choices are limited, residents

Table 7.2 Satisfaction with neighborhood, block and house by first language spoken

	Neighborhood	BBlock	HHouse	n
Vietnamese	8.0	8.0	8.1	91
English	8.3	8.4	8.4	354
Somali	9.3	9.2	9.1	19
Other	8.5	9.1	9.2	21
Cambodian	9.1	9.3	9.8	21

For neighborhood satisfaction, $F=2.8$ and $p=.025$. For block satisfaction, $F=3.47$ and $p=.008$. For house satisfaction, $F=3.837$ and $p=.004$.

will generally determine that their living situation suffices in meeting basic needs. Consequently, what may *appear to be* residential satisfaction, may simply be an adjustment to the inadequacies in one's housing, and remaining in place may not necessarily mean an enthusiasm for that place (see also Vale 1997). There is some evidence of this adaptability in this housing community:

> In general, we Vietnamese feel satisfied with what we have. Let's say, if we make $10,000 a month, it's enough to spend. But if we make $1,000 a month, it's still enough for us to spend. Even if we make only $500 a month, we still have 2 meals a day. For the Vietnamese, in general, we follow our ancestors' advice 'If we live in a tube, we become thin and long. If we live in a gourd bottle, we become big and short.' That's why we can adapt ourselves well to different circumstances.

However, people's tolerance of problems does not negate the positive dynamics and experiences of the community or the sense of well-being that it provided residents. In fact, it underscores how sense of place is "necessarily tentative and contingent, particularistic, dynamic, and potentially contradictory" (Butz and Eyles 1997, p. 6). Still, the preponderance of data from residents' perspective shows a very positive view of their community.

The Common Life

Residents' stories in focus groups and interviews reveal a thriving social community with a long history (Manzo and Kleit, forthcoming). The extent and nature of the neighboring and mutual support illustrate what I have come to consider the "common project" of living (Manzo, Kleit and Couch 2008). Residents visited one another and socialized (74 per cent), shared food (57 per cent), helped each other with errands (38 per cent), borrowed small items from each other (34 per cent) and watched each other's children (21 per cent). They took out each other's garbage, watched out for their neighbors' house or car when they were away (36 per cent), and helped one another with repairs (21 per cent). Over half of interviewees (53 per cent) said that they have relied or could rely on a neighbor in case of an emergency (Manzo, Kleit and Couch 2008). These are the very things that scholars claim are key ingredients for a successful community anywhere.

Interviewees periodically relied on their neighbors to give them rides, especially if they had limited mobility – either they had no car or it was broken, or they had physical difficulty getting around. One resident described how he helped a neighbor in this way:

> We had a gentleman here across the street I used to help all the time, but he died. Yeah. He had a stroke is why he had a wheelchair, but it was motorized, and he got around. I think he enjoyed going out, 'cause he went out almost every single day. Zooming down the hill in his chair. But there were times that he needed help, and I helped him. He needed help getting his drivers license. Took him down there to get his emission test. And one day, I carried his cat all the way to the vet.

Food was a particularly important community currency. It was something that people shared, having barbeques and holiday dinners together. People also sent food to one another, especially if a neighbor was ill, frail or elderly: ("This one lady, every weekend, she would cook and send me and my son a big old dinner"). Borrowing food was not uncommon either (Yes, yes! Maybe you run out of rice sometimes!").

There was a great deal of mutual support among neighbors despite cultural difference and language barriers. In fact, the cultural diversity among residents in the community was cited as a positive dimension by most interviewees (77 per cent). In describing the community, one resident explained, "We are the fruitbowl." As another resident put it, "When I want to make a friend, I make a friend, I don't care who." The fact that social support was so rich and extended beyond language and cultural distinctions demonstrates that this was not a socially isolated community; rather, there were opportunities for social engagement that many other neighborhoods do not provide.

Relocation Reactions and Concerns

Given these attachments, residents had trepidations about relocating. The vast majority of residents simply did not want to leave. As one resident put it:

> I was just so happy. I got the place, and it was where I wanted to be … I did not want to move at all. My mother passed away in that house … and I felt closer there [to] her being there. And I didn't really want to move from there because of that. And it feels like I left her behind.

Only 13 per cent of all household heads (n=507) were positive about the move (Manzo, Kleit and Couch 2008). Not surprisingly, residents' attitudes about moving were influenced by the outcomes that they anticipated, either positive or negative. Those who said something positive about the relocation thought that they had something to gain from moving, like better housing, while those who were reluctant to move focused more on what might be lost. Interestingly, the single-most commonly stated aspect of what residents would miss the most was their neighbors, although fully 63 per cent also mentioned some aspect of their physical housing as what they will miss most (location, having one's own yard). Together the social and physical features made this place home for people, and that was what they were ultimately fearing losing. As one resident explained, "What I want to do is enjoy every minute that I've got left because I really enjoyed living here." Another commented, "I'm going to be sad to leave because I made this my home, you know."

Concerns about moving fell into three categories: logistics of the physical move (packing and lifting for example), the cost of the move and their ability to re-establish their household after the move. The challenge of packing both logistically and emotionally emerged in the English language focus group. Here we see not only a severing of place attachments and social ties, but a severing of attachments to valued possession that served as meaningful markers of events through their lifecourse:

> P7: So that's where the stress picks up at. Especially when you're downsizing.

P3: That is what I am dealing with.

P4: You have to close your eyes and get rid of stuff.

P3: Yeah, well, how do you get rid of it?

P4: You close your eyes and get rid of it.

P6: It's hard.

P7: I've found myself throwing away birthday cards, anniversary cards …

P4: That's what I'm saying, you just close your eyes and you get rid of it …

In the end, whether residents were ready or not, they had to relocate. While the Housing Authority put an entire multi-lingual community support services staff in place to facilitate this process, the time and context in which this particular community thrived is now passed, and we can only hope that residents' housing, health and well being remain stable or improve as they relocate and establish themselves elsewhere. Follow up on this community is essential.

Conclusion

Although the negative impact of disruptions to place attachment on well-being is well established, housing programs and policies in the U.S. displace tens of thousands of low-income people and the full impacts of these are still unknown (Manzo, Kleit and Couch, 2008). To be sure, communities redeveloped through the HOPE VI program have their problems, both social and physical, but just like residents of other public housing communities, residents regard themselves as living in "communities with problems" rather than "problem communities" (Vale 1997, p. 173). This suggests that even when residents encounter problems in their housing, it does not fully define their lived experience of place (Manzo, Kleit and Couch 2008).

The implications of disruptions to place attachments for the health and well being of public housing residents must be better understood. Indeed, the only major national study of outcomes designed to address where residents move and how redevelopment affected their well being reveals that poor health is the biggest challenge for HOPE VI residents (Popkin 2007). That study shows that the poor health of residents had intensified over time, after residents relocated. Admittedly, without a true comparison group, it is difficult to know what might be attributable to the effectives of involuntary relocation, but research provides clear evidence that health issues are a major concern for HOPE VI residents (Popkin 2007).

At the time of this writing, the HOPE VI program is up for reauthorization, so how this program proceeds and is possibly reshaped is of considerable importance. In her testimony to the US Senate Subcommittee on Housing, Transportation and Community Development, Sue Popkin, a leading HOPE VI researcher, argued for more intensive medical services and supports for residents facing involuntary displacement, and a more holistic multifaceted approach that addresses the many factors that influence health and well being (Popkin 2007). Notably, she advocated for two years of case management and support to ensure successful transition, and for local Housing Authorities to coordinate with health providers and provide support throughout the relocation process.

Just as the above research reveals a connection between housing and health, other research shows a clear relationship between social connectedness, mutual support and health and well-being. Moreover, research specifically on the meaning of place shows the critical role of place and place attachments in well-being. Rootedness and a sense of belonging – indeed the larger construct of sense of place – play a fundamental role in providing stability, security and support in people's lives. These are the critical ingredients for psychological well-being, and they are nowhere more needed than in places of poverty. By giving us a solid existential ground on which to base our lives, reach our goals and live more meaningful lives, a sense of place can offer the very ontological security essential to a good quality of life.

This chapter sought to examine and discuss these issues, providing an inside perspective of residents' experiences of one public housing site as a way to illustrate the value and meaning such sites can have, and to raise questions about the implications of displacement on residents of socially well-functioning communities. It is too soon to tell what the full outcomes are for the residents of the site in question; follow-up research must be conducted. Whether any negative effects of displacing original residents will be outweighed over time by the benefits reaped by returning and new residents of the site remains an empirical question. We must also explore ways to improve the life circumstances for public housing residents without causing displacement that may be unnecessary and unhealthy in the long run. When determining the cost-effectiveness of redevelopment, the social and emotional costs to health and well-being must be factored into the equation.

Place attachments and sense of place may be highly private, but they are nevertheless grounded in a communicatively rationalized life world (Butz and Eyles 1997, p. 6). This gets particularly complicated in the case of public housing communities where we see competing world views. On the one hand, the residents themselves view such places as supportive and positive, yet some policymakers portray them as distressed and untenable. This underscores the ideological component of sense of place, which views sense of place as an intersubjectively shared lifeworld which people use as a basis to determine one another's validity claims (Butz and Eyles 1997). In the case of public housing redevelopment programs like HOPE VI, an understanding of this ideological component is helpful for a better understanding of both the experience of place and displacement. Here, we must determine the validity claims of the rhetoric of severe distress by juxtaposing them against residents' lived experience of place. Whose views prevail and who pays the cost?

The rhetoric of place can and has been used in service of certain social agendas. Some have argued that the rhetoric of severe distress has been used to justify a demolition approach to improving public housing and deconcentrating poverty, echoing an age-old approach that blames inner-city problems on the alleged social pathology of the poor (for more on this larger issue see especially Crump 2002). Additionally, the distress discourse has conflated the built and social aspects of public housing. In contrast to the images of a distressed community portrayed literally and figuratively in policy reports, the people I have met and the stories that they've told reveal a viable community where individually and collectively they were able to carry out their personal projects and share in the common project of living. Their experiences of place stand in sharp contrast to the life experiences implied in the sketch on the cover of the Commission's Final Report on Severely Distressed Public Housing as described at the beginning of this chapter.

Good neighborhoods are not simple achievements nor are they merely a matter of outward physical or economic improvements (Manzo and Perkins 2006). While comprehensive efforts to revitalize the human and physical fabric of declining neighborhoods are routinely advocated and implemented, proponents seldom focus on the role of place attachment and subjective assessments of the lived experience of place. Yet these are an essential part of our sense of well being. We must understand them better if we are to not perpetuate programs and policies that may threaten the well being of an already vulnerable population.

References

Altman, I. and Low, S. (1992), *Place Attachment* (New York: Plenum Press).

Berkman, L. and Glass, T. (2000), Social Integration, Social Networks, Social Support, and Health, in L. Berkman and I. Kawachi (eds). *Social Epidemiology* (Oxford: Oxford University Press).

Berkman, L. and Kawachi, I. (2000), *Social Epidemiology* (Oxford, UK: Oxford University Press).

Brown, B. and Perkins, D. (1992), Disruptions in place attachment, in I. Altman and S. Low (eds). *Place Attachment* (New York: Plenum Press).

Brown, B., Perkins, D. and Brown, G. (2003), Place Attachment in a Revitalizing Neighborhood: Individual and Block Levels of Analysis, *Journal of Environmental Psychology*, 23:3, 259–271.

Butz, D. and Eyles, J. (1997), Reconceptualising Senses of Place: Social Relations, Ideology and Ecology. *Geografiska Annaler*, 79B:1, 1–25.

Cresswell, T. (1996), *In Place/Out of Place: Geography, Ideology and Transgression* (Minneapolis, MI: University of Minnesota Press).

Crump, J. (2002), Deconcentration by Demolition: Public Housing, Poverty, and Urban Policy, *Environment and Planning D*, 20: 581–596.

Cunningham, M., Popkin, S. and Burt, M. (2005), Public Housing Transformation and the "Hard to House." A Roof Over Their Heads, Brief 9 (Washington DC: The Urban Institute).

Dunn, J.R. (2002), Housing and inequalities in health: A study of socioeconomic dimensions of housing and self reported health from a survey of Vancouver residents, *Journal of Epidemiology and Community Health*, 56, 671–681.

Easthope, H.V. (2004), A Place Called Home. *Housing, Theory and Society*, 21, 128–138.

Eyles, J. (1985), *Senses of Place* (Warrington: Silverbrook Press).

Farnum, J., Hall, T. and Kruger, L.E. (November 2005), Sense of Place in Natural Resource Recreation and Tourism: An Evaluation and Assessment of Research Findings. General Technical Report PNW-GTR-660.

Fried, M. (1963), 'Grieving for a Lost Home', in L. Duhl (ed.), *The Urban Condition: People and Policy in the Metropolis* (New York: Basic Books).

—— (2000), Continuities and Discontinuities of Place. *Journal of Environmental Psychology*, 20: 193–205.

Fullilove, M. (2004), *Root Shock: How Tearing up City Neighborhoods Hurts America, and What We Can Do About It* (New York: Ballentine Books).

Gans, H. (1962), *The Urban Villagers* (Glencoe, Il: Free Press).

Harris, P. B., Werner, C.M., Brown, B.B. and Ingebritsen, D. (1995), Relocation and Privacy Regulation: A Cross-Cultural Analysis, *Journal of Environmental Psychology*, 15: 311–320.

Hartmann, H. (1964), *Essays on Ego Psychology* (New York: International Universities Press).

Hayden, D. (1997), *The Power of Place: Urban Landscapes as Public History*, 2nd edn. (Cambridge: The MIT Press).

Hummon, D.M. (1992), Community Attachment: Local Sentiment and Sense of Place, in I. Altman and S.M. Low (eds). *Place Attachment* (New York: Plenum Press).

Kawachi, I. and Berkman, L. (2000), Social Cohesion, Social Capital, and Health, in L. Berkman and I. Kawachi (eds). *Social Epidemiology* (Oxford: Oxford University Press).

Kleit, R.G. and Manzo, L. (2006), To Move or Not to Move: Relationships to Place and Relocation Choices in HOPE VI, *Housing Policy Debate*, 17:2, 271–308.

Kuentzel, Walter F. (2000), Self-identity, Modernity and the Rational Actor in Leisure Research. *Journal of Leisure Research*, 32:1, 87–92.

Levy, D. (2005), Tending Home: Residents' Ambivalent Responses to Relocation from Public Housing Developments. Paper presented at the Urban Affairs Association conference Salt Lake City, UT.

Manjarrez, C. Popkin, S. and Guernsey, E. (June 2007), Poor Health: Adding Insult to Injury for HOPE VI Families. HOPE VI: Where Do We Go From Here? Brief No 5 (Washington DC: The Urban Institute).

Manzo, L. (2005), For Better or Worse: Exploring the Multiple Dimensions of Place Meaning. *Journal of Environmental Psychology*, 25:1, 67–86.

—— (2003), Beyond House and Haven: Toward a Revisioning of Place Attachment. *Journal of Environmental Psychology*, 23:1, 47–61.

Manzo, L.C., Kleit, R.G. and Couch, D. (2008), "Moving Once is Like Having your House on Fire Three Times:" The Experience of Place and Displacement Among Residents of a Public Housing Site, *Urban Studies* 45(9), 1855–1878.

Manzo, L.C. and Perkins, D.D. (2006), Finding Common Ground: The Importance of Place Attachment to Community Participation and Development, *Journal of Planning Literature*, 20:4, 335–350.

McCarty, M. (2005), *HOPE VI: Background, Funding, and Issues*, Order Code RL32236. Congressional Research Service, The Library of Congress, Washington, DC.

McGillivray, M. (2007), *Human Well-being: Concept and Measurement* (New York: Palgrave Macmillan).

Newman, K. and Wyly, E.K. (2006), The Right to Stay Put, Revisited: Gentrification and Resistance to Displacement in New York City, *Urban Studies*, 43:1, 23–57.

Popkin, S. (June 20, 2007), Testimony before the U.S. Senate Subcommittee on Housing, Transportation, and Community Development, and the U.S. House Subcommittee on Housing and Community Opportunity. Prepared for the Hearing on S. 829 HOPE VI Improvement and Reauthorization Act.

Relph, E. (1976), *Place and Placelessness* (London: Pion).

—— (1985), Geographical Experiences and Being-in-the-World: The Phenomenological Origins of Geography, in D. Seamon and R. Mugerauer (eds), *Dwelling, Place and Environment* (New York: Columbia University Press).

Sen, A.K. (1993), Capability and well-being, in M. Nussbaum and A. Sen (eds), *The Quality of Life* (Oxford: Clarendon Press for UNU-WIDER).

Shaw, M. (2004), Housing and Public Health, *Annual Review of Public Health*, 25: 397–418.

Simms, E.M. (2004), Urban Renewal and the Destruction of African-American Neighborhoods, *Environmental and Architectural Phenomenology*, 16:1, 3–6.

Steele, F. (1981), *The Sense of Place* (Boston, MA: CBI Publishing Company, Inc.).

Tuan, Y.F. (1976), Humanistic Geography, *Annals of the Association of American Geographers* 66: 206–276.

—— (1977), *Space and Place: The Perspective of Experience* (Minneapolis: University of Minnesota Press).

Vale, L.J. (1997), 'Empathological Places: Residents' Ambivalence toward Remaining in Public Housing', *Journal of Planning Education and Research*, 16: 159–75.

—— (2002), *Reclaiming Public Housing: A Half Century of Struggle in Three Public Neighborhoods* (Cambridge, Massachusetts: Harvard University Press).

Varady, D.P. and Walker, C. (2000), Vouchering out Distressed Subsidized Developments: Does Moving Lead to Improvements in Housing and Neighborhood conditions? *Housing Policy Debate*, 11: 115–162.

Ventakesh, S. (2000), *American Project: The Rise and Fall of a Modern Ghetto* (Cambridge, MA: Harvard University Press).

Wallenius, M. (1999), Personal Projects in Everyday Places: Perceived Supportiveness of the Environment and Psychological Well-Being, *Journal of Environmental Psychology*, 19:2, 131–143.

Chapter 8

Place, Leisure, and Well-being

Daniel R. Williams and Michael E. Patterson

Place and Leisure in Population Health

The Population Health Model (PHM) (Hamilton and Bhatti 1996) brings much needed attention to the critical role of environmental and community variables in the promotion of physical and mental health. However, it appears to neglect two facets integral to what is otherwise a rich contextual approach to health promotion. One facet, *place*, is the binding theme of this book and will be addressed accordingly. The other facet, *leisure* (and in particular nature-based recreation and tourism), is the specific substantive focus of this chapter. Few would seriously dispute that the quality of the environment plays a critical role in quality of life in all parts of the world in terms of providing basic material sustenance for life (food, shelter, etc.). Likewise, leisure and free-time activities are familiar venues for human development and quality of life. However, in the realm of public policy, with its inevitable bias towards tangible (demonstrable) outcomes, the role the environment (and more specifically relationships to place) in supporting the less tangible social, emotional, and spiritual dimensions of life quality – dimensions often manifest within the context of leisure and free-time pursuits – is easily overlooked.

The goal of this volume is to highlight the way the human/social geographic construct of *place* contributes to well-being and quality of life, with this chapter focused on relationships to the kinds of places that provide venues for leisure, recreation, and tourism (henceforth *leisure* will be used as a collective term for leisure, recreation, and tourism). In the health policy arena decision-makers generally want to know how the environment (variously defined and conceptualized) contributes (or possibly inhibits) human well-being and quality of life (also variously defined). Not unlike the population health approach, an important effort to frame these questions (albeit with an environmental policy focus) came out of the *Millennium Ecosystem Assessment (MEA)* project launched by the United Nations (Alcamo et al. 2003). This assessment involved over 1300 scientists from nearly 100 nations directed by an advisory board representing various international scientific and non-governmental organizations. Their aim was to address the needs of decision-makers for scientific information on the links between ecosystem change and human well-being.

Within the MEA framework the penultimate state of well-being is defined as freedom of choice and action or "being able to achieve what a person values doing or being" (Alcamo et al. 2003, 74), which is seen as dependent on other components of well-being namely: security, material needs, physical and emotional health, and social relationships. Though not explicitly recognized within the MEA, this description of well-being as freedom of choice and action mirrors the classic definitions of leisure grounded in Western thought. Accordingly, leisure may be viewed as both means and ends, with the MEA focused on leisure as an end. As an end "the time and activities spent in leisure may be profound expressions of culture and may be the most important part of life for finding relaxation, happiness and self-fulfillment" (Jackson 2006, 11). At the same time leisure is often seen as a means for achieving other goals at both individual and collective levels including physical and mental health, economic survival, human development and well-being, and environmental quality.

The MEA framework builds on the concept of ecosystem services taken from the emerging field of ecological economics to identify the major components of the environment that support this multidimensional concept of human well-being (Figure 8.1).

The various classes of ecosystem services include: supporting services such as soil formation and nutrient cycling; regulating processes such as climate, water purification, and waste treatment; provisioning services such as food, water, genetic resources, and raw materials production; and recreation and cultural services such as aesthetic, artistic, educational, spiritual, heritage, recreation, and scientific goods and services.

The overriding premise of the MEA – that the condition of the environment makes a vast and essential contribution to the quality of life and therefore protecting ecosystem services is paramount to ensuring human well-being – is consistent with a population health approach to well-being. Like the PHM, there are some

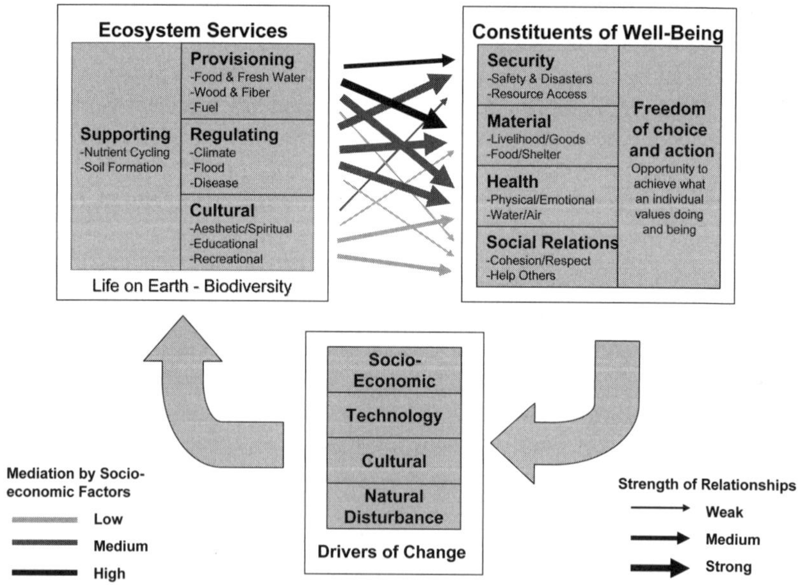

Figure 8.1 Ecosystem services and human well-being

Source: Alcamo, J. et al. (2003).

conceptual gaps stemming from its origins in biological as opposed to social theory. Reflecting the dominance of ecologically oriented scientists in the assessment, the MEA model implausibly characterizes cultural services as weak in their contribution to overall well-being compared to other ecosystem services, as well as being subject to relatively low levels of mediation by socio-economic factors. However, cultural services are somewhat unique among the classes of ecosystem services in that they are socially derivative of other ecosystem services and in that way are already mediated through human/cultural systems. In other words, the impact and mediation of cultural services is built into the construct itself as an end served by the other ecosystem services.

A key aim of this chapter is to illustrate what the concept of place brings to the environmental/contextual side of these models and what leisure or recreation brings to the well-being side. What we hope to do here is to: (1) show how a place approach to environmental research can be distinguished from other efforts to model environmental experience and quality of life and (2) illustrate how our own work on place has evolved toward the latter and what we think this work suggests about place and well-being in a nature-based leisure context.

Locating Place and Leisure within the Nexus of Environmental Well-being

The concepts of both place and leisure are typically peripheral considerations in theorizing about well-being and quality of life. Yet going back to at least Fried's (1963) classic study of grieving for homes lost to well-intended urban renewal projects, urban and regional planners have grappled with the notion of place as distinct from environment and its potential influence on well-being and quality of life. Likewise in ascribing "sentiment" and "symbolic" properties to urban spaces, Firey (1945, 140) challenged the dominance of a utilitarian or commodified view of the environment that presumed "the only possible relationship that locational activities may bear to space is an economic one." Firey and Fried were in the vanguard, advancing the idea that people could develop sentiments and attachments toward places that transcended the more stringent assessments of urban planners focused on the ostensibly objective environmental properties of spaces (the word *sentiment* itself connotes an attitude conditioned more by feeling than reason).

At about the same time Fried and others were contemplating the idea of place in the urban condition, academics and policy makers in the USA were describing the public's lack of access to nature and outdoor recreation as a national crisis. It was so urgent a problem the federal government organized a prestigious national commission to investigate the matter thoroughly. Established in 1958, the Outdoor Recreation Resources Review Commission produced a 27-volume report that sparked two decades of landmark environmental legislation in the USA and inspired a worldwide environmental movement. Today that same sense of urgency is barely palpable, dampened according to Louv (2005), by the rising popularity of electronic amusements and an overprotective and litigious culture worried about dangers to children lurking behind every tree. While various urban crises of the 1960s instigated a train of commentary pointing to weakened ties to place and reduced and devalued access to nature and outdoor recreation as grave threats to quality of life, they remain under-examined aspects of most models of well-being.

The key to understanding how place informs leisure and thereby well-being is to differentiate it from two alternative (and often dominant) paradigms in the study of environmental experience. First, the *adaptive* paradigm is a common (and very medical) approach that starts with the assumption that biological and psychological survival motivates behavior. Typical adaptive responses examined include how people cope with stressful environments and how certain environments or environmental features serve restorative or therapeutic needs. Research on how organisms cope with stressful environments typically looks for direct dose-response linkages between specific environmental stimuli (e.g., sound or temperature), and psychological functioning or well-being. Generalizing from community noise studies for example, a dose-response model was used to explain the negative impact of aircraft noise on outdoor recreation experiences (Mace et al. 1999). Leisure is recognized as both a context within which people find opportunities to cope with daily stressors (Hull and Michaels 1995), as well as a context in which people sometimes encounter stress (e.g., crowding, environmental threats) and must adapt or cope in some way with stressors in the recreation environment, which may diminish the stress relieving value of recreation (Iwasaki and Schneider 2003). Whereas the stress model portrays leisure as a way to overcome negative environmental conditions to maintain health and well-being, some research goes further to suggest certain environmental features and settings have an intrinsic capacity to promote physical healing and mental restoration (Hartig and Staats 2003). Accordingly, human responses to the environment are better adapted to natural stimuli, and therefore exposure to nature promotes well-being (Ulrich 1993).

The adaptive model of the environment-experience relationship has been very influential in establishing important health benefits of leisure (Driver et al. 1991). However, while the quality of the evidence is strong and convincing, by conceptualizing the environment as a natural (as opposed to socially defined) phenomenon, studies following the adaptive paradigm tend not to address the larger context of place including economic, social, and political forces that structure environmental conditions and distributions of power to access and regulate these conditions within society. In the adaptive paradigm the environment has inherent capacities to promote well-being but, like the MEA, tends to assume little mediation by social, economic, and political conditions.

Second, the goal-directed or *utility* paradigm starts with the assumption that well-being derives from opportunities in the environment to satisfy specific social or intra-personal needs. In the context of leisure the environment is interpreted as a "setting for action" possessing the characteristics necessary for the pursuit of specific desired activities such as rivers for rafting, snow for skiing, paths for walking, etc. People are viewed as rational decision-makers, satisfying individual needs, rather than responders to the biological imperatives of the adaptive model. Research examines how people evaluate environmental attributes in arriving at a decision or action, which makes this approach well-suited to the instrumental and rational traditions of environmental and program planning (Williams and Patterson 1996). The specific goals people associate with various forms of leisure participation are generally learned (as opposed to innate) and pursued in situations (times and places) that people perceive to offer good opportunities for their satisfaction. Because goal preferences are learned they are assumed to vary widely by personality, social background, culture, and geography. It is also the case, however, that people with similar backgrounds may still hold different preferences for environmental features to satisfy similar psychological goals and that different goals are often satisfied by similar environmental (setting) features (Knopf 1987).

Inherent in the utility paradigm is the notion that environmental features are theoretically interchangeable (i.e., substitutable) and even reproducible, so long as the replacement provides a similar combination of goal-fulfilling attributes. Psychological responses (e.g., satisfaction of leisure goals) are understood to be instrumentally dependent on specific properties of the environment. Evidence supporting this model is relatively strong for ecosystem or environmental services that are generic and homogeneous (e.g., recreation at an amusement park), but weak when applied to the environmental services that are more ambiguous (e.g., environmental awareness) or geographically unique (e.g., historical preservation). A key advantage of the utility paradigm is that it can be integrated with economic approaches to resource valuation (Loomis 2002) and allows for segmenting various social groups by environmental preferences to aid marketing and management (Mallou et al. 2004). The weaknesses of this model, however, are that it provides limited understanding of the socioeconomic and socio-cultural factors influencing the distribution of opportunities for individual goal attainment, reduces all environmental values to behavioral utilities, and generally ignores the symbolic construction (meaning) of the environment.

Both of these models provide strong support for the role of specific environmental qualities in contributing to health and well-being. Not unlike the PHM and the MEA, however, the adaptive and utility paradigms treat place as little more than a container of environmental features with well-being a more or less a direct affective response to those features. Within the *place* paradigm environmental preferences extend well beyond biological imperatives and individual goal-oriented opportunities. Accordingly, leisure experiences of a place are socially constructed within the cultural, historical, and geographical context of day-to-day life (Farnum et al. 2005; Williams and Patterson 1996). People are seen as social beings seeking out and creating meaning in the environment (Knopf 1987). From the place paradigm, any single environmental feature may be perceived from a variety of social or cultural perspectives. Wilderness may, for example, symbolize ancestral ways of life, spiritual contemplation, valued commodities, tourist experiences, or essential livelihood to different groups of people. Thus, the environment acquires varied and competing social and political meaning as a specific place associated over time with particular activities and groups (Lee 1972). As modern social relations have become more mobilized with sense of belonging and rootedness more diffused and fragmented across multiple places, leisure participation appears to offer people a way to negotiate multiple senses of place, home, and identity that enhance their sense of well-being (McIntyre et al. 2006). In addition, research is beginning to examine social and cultural differences in access to the economic and political resources necessary to define and direct the use of leisure settings; the basis of much inter-group conflict (Stokowski 2002).

Socio-cultural studies of place emphasize the way landscape features and settings are symbolically constructed as leisure places (Blake 2002; Stokowski 2002), both through the meanings ascribed to them by users, tourists, and local residents and by the intentions of designers, developers, and promotional and managing agencies (McCraken 1987; Mick and Buhl 1992). As a result leisure places are subject to complex, contested social processes in which various stakeholders struggle to manipulate and control place meanings, values, and uses (Williams and Van Patten 2006). These approaches are proving increasingly valuable to managers and policy makers as they try to balance the competing environmental priorities of diverse constituencies. Thus, for managing parks, protected areas, and tourism destinations, leisure can be a destabilizing force in the sustainability of landscapes and local culture or a potential vehicle for the their preservation (Williams 2001). While the quality of the evidence supporting the basic role of symbolic construction of leisure settings in maintaining and enhancing a sense of collective identity and community is substantial, this approach tends to overlook the role and value of individual meanings and setting specific attachments that users form with favorite leisure settings.

Place and Well-being in Nature-based Leisure

Psychological well-being associated with leisure has long been a core research question within the professional practice of natural resource management. Through much of the 1970s and 1980s the work of Driver and associates championed a program of research to document and quantify the quality-of-life benefits of participation in outdoor pursuits (Driver et al. 1991). However, this model focused on the act of recreating largely independent of the place it occurred. Following the utility paradigm, preferences for environmental features were considered fungible[1] elements in a recreation decision-making process; absent was any consideration of the meaning the user attached to the specific places they were found. Much like Firey and Fried who elevated the study of place in urban planning, a few investigators studying recreation, nature, and park use developed parallel interests in the experience of place. The basic idea they began to explore was that people often value their relationships to natural landscapes and outdoor places not merely because they were useful settings for pursuing desired outdoor recreation activities, but because the specific places involved conveyed a sense of individual identity and group affiliation. Drawing inspiration from Tuan's (1977) distinction between space perceived as a territory for an activity versus place as a locus for feelings of attachment and belonging, they noted that over time people accumulate meaning and form emotional ties to specific places (Schreyer et al. 1981), establish "social

1 No other word captures our intended meaning as well as the financial term, *fungible*. According to the *Random House Dictionary of the English Language*, fungible (especially as applied to goods) means "being of such nature or kind as to be freely exchangeable or replaceable, in whole or in part, for another of like nature or kind." The most fungible of all goods would be money itself.

definitions" (Lee 1972) and "feelings of possession" (Jacob and Schreyer 1980) about places, and seek out places where the norms of behavior and expressed values and lifestyles match their own (Lee 1972; Schreyer and Roggenbuck 1981).

Place Attachment

Building on this work Williams and colleagues initiated an effort in the late 1980s to develop a psychometric instrument that could measure the strength of place orientations for use in management-directed public surveys of recreation visitors to national parks, forests, and other wildland and tourist destinations (Williams and Vaske 2003).[2] Over the years this psychometric instrument has been employed, sometimes in modified form, in various contexts (Farnum et al. 2005) ranging from leisure places (Williams et al. 1992), second homes (Kaltenborn 1997), local communities (Vorkinn and Riese 2001), and birthplaces (Nanistova 1998). Given that leisure affords the individual considerable discretion to affiliate with places and activities of ones choosing (Haggard and Williams 1991), attachment to leisure places has typically focused on positive relationships, with leisure places serving as powerful symbols of individual identity. Applied to other kinds of places, particularly at a community level, a sense of attachment sometimes may be experienced negatively as a sense of entrapment and inability to leave (Manzo 2003). Still, the psychometric approach to place attachment appears particularly well suited to measuring the strength of personal emotional bonds (e.g., meaningfulness or sentiment) as opposed to the meanings themselves (e.g., the place as symbolic of one's values or beliefs). They help investigators identify the places that may indeed serve as powerful symbols without getting too deeply into the origin and nature of those symbols and their impact on well-being. To study the more elusive and amorphously multifaceted notion of sense of place and the individual and socio-cultural meanings that go with place, we have turned to more qualitative, interpretive methods. The focus has been to examine how place relationships are an important part of nature experiences and how they contribute meaning, stability, identity, and ultimately enhance well-being.

Experience and Meaning of Places

Recognizing some of the limits of the place attachment approach, we turned our attention more directly to meanings and sense of place. This shift stimulated deeper reflection on the ontological issues implied by the idea of place in the work of human geographers such as Tuan (1977) and Relph (1976) and drew our attention to parallels between their phenomenological views of place and those social-psychological models of well-being that emphasize transaction and meaning (Patterson and Williams 2005). Specifically, Omodei and Wearing (1990) grouped

2 The design of the scale built on Brown's (1987) suggestion of two forms of place attachment. One was Stokols and Shumaker's (1981) the concept of place dependence, which represented the importance of a place in providing features and conditions that support specific goals or desired activities. The other was Proshansky's (1978) concept of place identity, referring to the importance of a place in constructing and maintaining self-identity.

conceptual models of well-being into two broad classes: (1) telic or goal attainment models that view well-being as occurring when specific needs or goals are met, and (2) auto-telic or process oriented models that view well-being as arising directly from the nature of activity and from interaction with objects, places, and people rather than from attaining desired end states.

As noted earlier, until the 1990s goal-attainment models dominated research exploring the contribution of leisure to well-being. However, sense of place as conceived in human geography appeared to be more consistent with a process oriented model of well-being. Rather than beginning with a view of leisure participants as rationally processing information about objective attributes of a place in relation to specific goals obtained through participation in isolated activities or events with a definite beginning and end, they are viewed as participating in the ongoing enterprise of constructing a life and an identity (McCracken 1987). People are not seen as responding deterministically and/or passively to information that objectively exists in specific environmental attributes. Instead, meaning is viewed as an emergent property that is actualized through a transactional relationship between person and the place (Mick and Buhl 1992).

Ultimately, we came to believe that this more interpretive perspective on ontology (i.e., the view that meaning arises from a transactional process rather than resides inherently in places or their attributes) required an interpretive epistemology that was not reflected in the psychometric approach employed in both the dominant utility paradigm and the measurement of place attachment as summarized above. Indeed, much of the earlier work on place in human geography was from a phenomenological perspective, reflecting a qualitative empirical approach that is much closer to the hermeneutic approach we began to adopt (Patterson and Williams 2002).

Following this hermeneutic approach our first study explored leisure experiences while canoeing a slow moving, spring fed creek in a Florida wilderness area (Patterson et al. 1998). Rather than defining a predetermined operational model as we had done with the place attachment work, we started with experiential narratives (descriptions of the experience in participants' own words). This research yielded several insights relative to the origins of well-being and nature-based recreation. First it suggested that, in many cases, experiences had a storied nature best understood in terms of an emergent narrative rather than a set of expectations. That is, rather than a precise set of specific goals, many participants seemed motivated by a not very well-defined purpose of acquiring stories that ultimately enrich their lives. For example, challenge, which emerged as a salient dimension in virtually all the narratives, became not an end goal but a highly variable dimension describing the character of the experience. It was variable because the nature and meaning of "challenge" and the role it played in the experience varied across individuals.

A second insight was the dynamic nature of the emergent narratives as meaning often shifted over time. For example, in some cases those who experienced an intense challenge that had unpleasant aspects debated in their own narrative whether or not the experience was a positive one. Several participants ultimately came to see it in a positive light with a sense of achievement even though they would have liked to quit half way through. As another example, some participants who experienced the challenge less intensely initially complained about aspects of the setting that contributed to the degree of challenge (e.g., snags and blow down trees) but upon reflection and discussion came to realize that these were precisely what made the experience an enjoyable story to relive. Finally, for yet another kind of participant, challenge was more appropriately described as *a* defining characteristic of the experience which served as a key aspect in building an enduring relationship to the place that was important both to their identity and quality of life.

Thus a more interpretive approach did seem capable of yielding richer insights into participants' meanings or senses of place. Challenge, just one facet revealed in the experiential narratives, seemed to be a key dimension defining the character and meaning of the place to visitors. Further, the study suggested another important place relationship-based distinction among respondents. Some respondents truly seemed to be well described by the label "visitors" (individuals seeking to add to their quality of life through a single, isolated experience at this place). Others clearly had a more enduring relationship to place, one so strong that the term visitor might actually represent a mischaracterization of how the place relates to their well being. A subsequent study of motorized users on a whitewater river (the River of No Return) in Idaho provided an opportunity to explore this type of more enduring recreational relationship to natural places.

Also using in-depth interviews and an interpretive (hermeneutic) approach, this study provided a more comprehensive characterization of relationship to place, encompassing the significance and nature of the place bond, how lives of those with significant bonds were organized around the place, and place meanings. Four different types of bonds were reflected in the interviews. The first two reflected highly significant bonds corresponding to the two primary conceptualizations described above in the discussion of place attachment: place identity and place dependence. The other two represented comparatively low intensities of attachment.

Those respondents with deeply rooted place identity and place dependent bonds tended to organize their lives around recreational use of the river in multiple ways. First, the river provided both a setting and a community of users in which strongly bonded respondents actively constructed highly valued identities. This sense of identity was not merely a personal one. Respondents' relationships to the river were capable of promoting a broader family identity – a sense of connectedness that transcended generations and became an important link to the past and a valued heritage to be passed on to the next generation.

The high degree of investment in place in terms of self and family identity exhibited by those with deeply rooted bonds is mirrored by other forms of investments that are more tangible in nature and that have significant implications for other dimensions of life. Examples include choosing a life of lower wages to have the opportunity to recreate in this way in this place, making a long term family commitment to a particular form of recreation as a result of the significant financial investment motorized whitewater boating requires, and choosing an insurance company due to the adequacy of their coverage for their recreational boats. Additionally, many attached boaters joined a regional power boating club, not only for social opportunities but also as a means to establish an effective political voice to represent their place interests in the area planning process. Further, one of the clubs sought not only to establish a political means to protect their access interests but also to promote a use and stewardship ethic among its own members and motorized boaters in general.

The study of motorized whitewater boaters shows that people can form strong emotional bonds with nonresidential nature-based recreation places through place identity or place dependent processes that are so central to a sense of self that individuals come to organize their lives around these places in very tangible ways. These transactional relationships may generate meanings for a place that conflict with other meanings and must be resolved in a political process in which place-based bonds (this place as unique) may conflict with meanings arising through other sources (e.g., as national parks or protected areas). This conflict may significantly impact quality of life. Both the canoeist and motorized boater studies suggest that an interpretive epistemology employing descriptive narratives can yield significant insights into relationship to place and quality of life not attainable through quantitative psychometric approaches. Just as important, these approaches are capable of yielding insights into how policy and decision-making may affect well-being and quality of life.

Mobility and Multicentered Place Identities

As we have seen, place-based sentiments and symbolism associated with leisure and nature contact provide opportunities to establish and express individual identity, maintain a coherent self-narrative, and provide a sense of rootedness. However, these findings also suggest a paradox. Much social inquiry of the past three decades has emanated from critiques of globalization and the associated fragmentation, if not dissolution of self and identity (McIntyre et al. 2006; Giddens 1991). Yet, our work with leisure participants suggests that greater mobility also enables a wider search for thicker place meanings and deeper ties to place. This search can be likened to what Giddens (1991) calls the "reflexive project of the self" and, in particular, exemplifies various emerging social mechanisms he describes for crafting coherent identities under the turbulent, fragmented conditions of late modernity. For Giddens, constructing an identity under these conditions is a more deliberate and cognitively demanding project because it must be accomplished amid a greater diversity of lifestyle options, competing sources of authority and expertise, and extensive access to multiple localities thoroughly penetrated by distant global influences.

To better understand how people might use leisure to negotiate this tension between freedom/mobility and attachment to place, Williams and colleagues initiated a series of studies examining the use and meaning of second homes (Williams and Kaltenborn 1999; Williams and Van Patten 2006; McIntyre et al. 2006) as a strategy for organizing and maintaining coherent identity narratives amid the disorienting and fragmenting din of modernity. A common theme in interviews with second homeowners involves escaping modernity by seeking refuge in nature. In many cases the second home affords greater access to nature, not only because of its typically rural location, but also by virtue of the spaciousness of most cottage developments relative to suburban living. It also affords an immersion in the rhythms of nature with a corresponding shift in awareness and a mental refocus on simpler ways of living.

Another theme in interviews with second homeowners is continuity and rootedness. To the extent that modernity thins the primary home of meaning, people seem to cultivate alternative mythic places such as second homes or favorite vacation spots to recreate a seemingly thicker place of attachment, identity, continuity, and tradition otherwise undermined by modern lifestyles. Second homes provide centers of meaning or an emotional base across the life course even as people relocate their so-called permanent residence. Thus the second home is often the locus of important family memories and provides for a more consistent and enduring territorial identification for families across generations. However, these oft cited virtues are to some degree mitigated by a sense of obligation and burden that comes with maintaining a second home (e.g., yard work, repairs etc.).

Though cottage life offers a seemingly thicker place of identity, continuity, and tradition, there is also a contradiction in such efforts as suggested by Giddens' identity dilemma of unity versus fragmentation. As noted earlier organizing a coherent and secure sense of self in modern times must be accomplished amid an increasingly fragmented and puzzling diversity of options and possibilities. The second home is to some extent a modern expression of this dilemma as a concrete manifestation of a segmented, isolated self living in more than one place, but also needing to have an authentic, rooted identity somewhere (Williams and Kaltenborn 1999). It necessarily re-creates the segmented quality of modern identities in the form of separate places for organizing distinct aspects of a fragmented identity. It narrows and thins out the meaning of each "home" by focusing the meaning of each on a particular segment of life and it segments identity around phases in the life cycle with youth and retirement focused more on cottage life than working adulthood.

At the same time cottage life provides opportunities to create distinctive identities by incorporating elements from diverse settings into an integrated narrative that some have described as a "cosmopolitan identity" (Gustafson 2006). The cosmopolitan person is one who possesses the distinct knowledge and competence to handle cultural diversity and who draws unique strength from being at home in a variety of contexts. Much like Giddens, Bruner (1990) envisions self-identity as distributed through our personal undertakings and suggests that meaning in life is created reflexively through narratives, both as the stories we tell to our "self" and the stories we tell to others. While recognizing modernity's fragmenting and disorienting qualities, leisure and various forms of multiple dwelling constitute strategies people have available and draw on to assemble and narrate a coherent yet multicentered identity.

Conclusions

Our studies of sentiments and symbolism associated with places people seek out for leisure grew out of a critique of the utilitarian ethos that has long dominated the professions of forestry and natural resource management, an ethos reflected as well in the rational planning ideologies underlying other realms of practice such as urban development and health promotion. In contrast, our initial aim was to document the attachments people felt toward particular outdoor recreation places as a way to demonstrate that the benefits people sought from experiences in these places went well beyond the commodity or consumer metaphor embedded in need satisfaction theories widely used to explain outdoor recreation participation. Recognizing that measuring the intensity of attachment only scratched the surface of deeper questions concerning how people derive meaning from outdoor experiences, we began to employ interpretive epistemology to investigate the meanings people associated with single events or experiences of a place. From these studies we noted how, for many people, relationships to specific locales were a central organizing facet of their individual and group identities. At the same time we began to explore how people take advantage of modern mobility through leisure and construct ties to multiple places (e.g., second homes) as a modern way to anchor an identity in an age of globalization.

In trying to show how people indeed seek place-based meaning and identity through interactions with multiple distant locales, we took inspiration from those who criticized the disruptions to spatial relationships and experiences that followed in the wake of urbanization and globalization. This had more to do with a shared concern over the corrosive effects of narrow instrumental thinking on the modern conception of the meaning and value of places than the belief that mobility constituted a threat to traditional sources of identity and social cohesion. Like Gustafson (2001) and Lippard (1997) our work has tried to go beyond the simple opposition between mobility and attachment to demonstrate alternative ways people might create place-based meaning and identity in this late-modern age dominated by global spatial and social relations. Ties to place continue to be desired and necessary even in the globalized age. At least for some, this does not necessarily occur solely by sinking roots even deeper into local soil. Rather, the idea of multicentered place identities offers a more flexible notion of place-based well-being, one suited to modern (global) social relations.

Broad deterministic frameworks on human well-being such as the PHM and the MEA tend to favor biological, adaptive, and/or instrumental models of environmental effects in which the environment is conceived as a constellation of separable and generally substitutable (fungible) properties that have consequences for human health and well-being. Such an assumption goes back, for example, to 1960s urban renewal projects in which urban planners presumed that forcing people out of slums and into objectively superior housing would make their lives better (Fried 1963). In contrast the concept of place as developed across a number of environmental fields presumes that well-being involves, in addition, a sense of belonging or identification with an environment as a specific place and that places contribute to well-being in ways that are not easily managed or reproduced by the intentions of planners and designers. Moreover, this chapter has attempted to show that people benefit from a sense of involvement, belonging, and/or identification with places they use for leisure or free-time pursuits. From our studies we noted how, for many people, relationships to specific locales were a central organizing facet of their individual and group identities. At the same time we have sought to show how people take advantage of leisure and modern mobility to construct ties to multiple, often distant places (e.g., second homes) as a modern way to anchor an identity in an age of globalization. In the health promotion world, we need to recognize not only the more instrumental effects of environmental conditions and designs on well-being, but the needs of people to establish and maintain some control over their relationships to specific places that contribute to their sense of belonging and identity. Leisure, recreation, and tourism are important venues for building and maintaining such relationships to place.

References

Alcamo, J. and Bennett, E.M. (2003), *Ecosystems and Human Well-Being: A Report of the Conceptual Framework Working Group of the Millennium Ecosystem Assessment* (Washington: Island Press).

Blake, K.S. (2002), 'Colorado Fourteeners and the Nature of Place Identity', *Geographical Review* 92: 155–176.

Brown, B.B. (1987), 'Territoriality', in D. Stokols and I. Altman (eds), *Handbook of Environmental Psychology* (New York: Wiley).

Bruner, J. (1990), *Acts of Meaning* (Cambridge: Harvard University Press).

Driver, B.L. et al. (1991) (eds), *Benefits of Leisure* (Champaign, IL: Sagamore).

Farnum, J. et al. (2005), *Sense of Place in Natural Resource Recreation and Tourism: An Evaluation and Assessment of Research Findings*, PNW-GTR-660 (Portland, OR: USDA Forest Service, Pacific Northwest Research Station).

Firey, W. (1945), 'Sentiment and Symbolism as Ecological Variables', *American Sociological Review* 10: 140–148.

Fried, M. (1963), 'Grieving for a Lost Home', in L.J. Duhl (ed.), *The Urban Condition* (New York: Basic Books).

Giddens, A. (1991), *Modernity and Self-identity: Self and Society in the Late Modern Age* (Stanford: Stanford University Press).

Gustafson, P. (2006), 'Place Attachment and Mobility', in N. McIntyre et al. (eds), *Multiple Dwelling and Tourism: Negotiating Place, Home and Identity* (Cambridge: CABI).

Gustafson, P. (2001), 'Roots and Routes: Exploring the Relationship Between Place Attachment and Mobility', *Environment and Behavior* 33: 667–686.

Haggard, L.M. and Williams, D.R. (1991), 'Self-Identity Benefits of Leisure Activities', in B.L. Driver et al. (eds), *Benefits of Leisure* (State College, Pennsylvania: Venture Publishing).

Hamilton, N. and Bhatti, T. (1996), *Population Health Promotion: An Integrated Model of Population Health and Health Promotion* (Ottawa: Health Promotion Development Division, Health Canada).

Hartig, T. and Staats, H. (2003), 'Guest Editors' Introduction: Restorative Environments', *Journal of Environmental Psychology* 23: 103–108.

Hull, R.B. and Michael, S.E. (1995), 'Nature-Based Recreation, Mood Change, and Stress Restoration', *Leisure Sciences* 17: 1–14.

Iwasaki, Y. and Schneider, I.E. (2003), 'Leisure, Stress, and Coping: An Evolving Area of Inquiry', *Leisure Sciences* 25: 107–114.

Jacob, G.R. and Schreyer, R. (1980), 'Conflict in Outdoor Recreation: A Theoretical Perspective', *Journal of Leisure Research* 12: 368–380.

Jackson, E.L. (2006), 'Summary Principles about Leisure', in E.L. Jackson (ed.), *Leisure and the Quality of Life: Impacts on Social, Economic and Cultural Development* (Hangzhou: Zhejiang University Press).

Kaltenborn, B.P. (1997), 'Nature of Place Attachment: A Study Among Recreation Homeowners in Southern Norway', *Leisure Sciences* 19: 175–189.

Knopf, R.C. (1987), 'Human Behavior, Cognition, and Affect in the Natural Environment', in D. Stokols and I. Altman. (eds), *Handbook of Environmental Psychology* (New York: John Wiley).

Lee, R.G. (1972), 'The Social Definition of Outdoor Recreational Places', in W.R.J. Burch et al. (eds), *Social Behavior, Natural Resources and the Environment* (New York: Harper and Row).

Lippard, L.R. (1997), *The Lure of the Local: Sense of Place in a Multicentered Society* (New York: The New Press).

Loomis, J. (2002), 'Valuing Recreation and the Environment: Revealed Preference Methods in Theory and Practice', *Environmental and Resource Economics* 21: 104–105.

Louv, R. (2005), *Last Child in the Woods: Saving Our Children from Nature-Deficit Disorder* (Chapel Hill: Algonquin).

Mace, B.L. et al. (1999), 'Aesthetic, Affective, and Cognitive Effects of Noise on Natural Landscape Assessment', *Society and Natural Resources* 12: 225–242.

Mallou, J.V. et al. (2004), 'Segmentation of the Spanish Domestic Tourism Market', *Psicothema* 16: 76–83.

Manzo, L. (2003), 'Beyond House and Haven: Toward a Revisioning of Emotional Relationships with Places', *Journal of Environmental Psychology* 23: 47–62.

McCracken, G. (1987), 'Advertising: Meaning or Information?', *Advances in Consumer Research* 14: 121–124.

McIntyre, N. et al. (eds) (2006), *Multiple Dwelling and Tourism: Negotiating Place, Home and Identity* (Oxfordshire: CABI).

Mick, D.G. and Buhl, C. (1992), 'A Meaning-Based Model of Advertising Experiences', *Journal of Consumer Research* 19: 317–338.

Nanistova, E. (1998), 'The Dimensions of the Attachment to Birthplace and their Verification after the 40 Years Following Forced Relocation', *Sociologica* 30: 337–394.

Omodei, M.M. and Wearing, A.J. (1990), 'Need Satisfaction and Involvement in Personal Projects: Toward an Integrative Model of Subjective Well-Being', *Journal of Personality and Social Psychology* 59: 762–769.

Patterson, M.E. and Williams, D.R. (2005), 'Maintaining Research Traditions on Place: Diversity of Thought and Scientific Progress', *Journal of Environmental Psychology* 25: 361–380.

—— (2002), *Collecting and Analyzing Qualitative Data: Hermeneutic Principles, Methods, and Case Examples* (Champaign: Sagamore).

Patterson, M.E. et al. (1998), 'An Hermeneutic Approach to Studying the Nature of Wilderness Experiences', *Journal of Leisure Research* 330: 423–452.

Proshansky, H.M. (1978), 'The City and Self-Identity', *Environment and Behavior* 10: 147–169.

Relph, E. (1976), *Place and Placelessness* (London: Pion Limited).

Schreyer, R. et al. (1981), 'Environmental Meaning as a Determinant of Spatial Behavior in Recreation', *Proceedings of the Applied Geography Conferences* 4: 294–300.

Schreyer, R. and Roggenbuck, J.W. (1981), 'Visitor Images of National Parks: The Influence of Social Definitions of Places on Perceptions and Behavior', in D. Lime and D. Field (eds), *Some Recent Products of River Recreation Research*, GTR NC-63 (St. Paul: USDA Forest Service, North Central Research Station).

Stokols, D. and Shumaker, S.A. (1981), 'People in Places: A Transactional View of Settings', in D. Harvey (ed.), *Cognition, Social Behavior, and the Environment* (Hillsdale: Erlbaum).

Stokowski, P.A. (2002), 'Languages of Place and Discourses of Power: Constructing New Senses of Place', *Journal of Leisure Research* 34: 368–382.

Tuan, Y-F. (1977), *Space and Place: The Perspective of Experience* (Minneapolis: University of Minnesota Press).

Ulrich, R.S. (1993), 'Biophilia, Biophobia, and Natural Landscapes', in S. Kellert and E.O. Wilson (eds), *The Biophilia Hypothesis* (Washington: Island Press).

Vorkinn, M. and Riese, H. (2001), 'Environmental Concern in a Local Context: The Significance of Place Attachment', *Environment and Behavior* 33: 249–363.

Williams, D.R. (2001), 'Sustainability and Public Access to Nature: Contesting the Right to Roam', *Journal of Sustainable Tourism* 9: 361–371.

Williams, D.R. and Kaltenborn, B.P. (1999), 'Leisure Places and Modernity: The Use and Meaning of Recreational Cottages in Norway and the USA', in D. Crouch (ed.), *Leisure Practices and Geographic Knowledge* (London: Routledge).

Williams, D.R. and Patterson, M.E. (1996), 'Environmental Meaning and Ecosystem Management: Perspectives from Environmental Psychology and Human Geography', *Society and Natural Resources* 9: 507–521.

Williams, D.R. and Schreyer, R. (1981), 'Characterizing the Person-Environment Interaction for Recreation Resource Planning', *Proceedings of the Applied Geography Conferences* 4: 262–271.

Williams, D.R. and Van Patten, S.R. (2006), 'Home *and* away? Creating Identities and Sustaining Places in a Multi-Centered World', in N. McIntyre et al. (eds), *Multiple Dwelling and Tourism: Negotiating Place, Home and Identity* (Cambridge, MA: CABI).

Williams, D.R. and Vaske, J.J. (2003), 'The Measurement of Place Attachment: Validity and Generalizability of a Psychometric Approach', *Forest Science* 49: 830–840.

Williams, D.R. et al. (1992), 'Beyond the Commodity Metaphor: Examining Emotional and Symbolic Attachment to Place', *Leisure Sciences* 14: 29–46.

Chapter 9

Sense of Place, Well-being and Migration among Young People in Sarajevo

Carles Carreras[1]

This chapter was conceived on the occasion of the 20th anniversary of the publication of the book written by John Eyles, during his last visit to Spain. The senses of place analysis is seen as a useful concept that significantly helps to explain the general emigration trend of many young people in some metropolitan areas of an undelimited and undefined South.[2]

This chapter is based on the results of long and intermittent research project based on fieldwork carried out in Sarajevo (Bosnia)[3] with different generations of university students, between 1997 and 2006. Respondents were generally well educated, and included teachers, students, researchers and international participants. Recognizing the bias of the sample, the qualitative approach encompassed the use of field and floating observation,[4] mental maps analysis, and many in-depth interviews. The main issues of the work have been reinterpreted in order to connect senses of place and ideas of welfare and well-being. Some results of the research are already in the publication progress;[5] others are in their last steps.

---

1 The author must thank his colleague Sergio Moreno for his useful comments and suggestions to the earlier versions of this chapter.

2 There is not a correct denomination for the whole of different countries that both from quantitative as well as qualitative points of view are clearly below the so-called western way of life. Traditional terms, like Developing countries or Third World become too general if not a terrible mistake (many countries spend years waiting for its own development). We prefer here to use the also general and undelimited geographical notion of South that implies not only socio-economic differences, but also cultural, in the sense of the *carrollian* image of the other side of the mirror.

3 We have also undertaken very similar research in the Brazilian city of Sâo Paulo since 1987 and in the Egyptian city of Alexandria since 2004. The different cultural contexts prohibit us to integrate them. But it must be noted that many problems appears to be very similar among young people everywhere.

4 Floating observation is a non directed form of observation, as been defined by the French anthropologist Colette Petonnet in 1982.

5 "Periferie urbane, esclusione sociale e istituzionalizzazione della delinquenza. I casi di Rio de Janeiro e San Paolo nel Brasile" in Gribaudi, G. and Somella, R. *Violenze* urbane, Bollati e Boringheri, Torino (in press); and with Moreno, S. "Los procesos de modernización en Sarajevo. La incierta dirección de la flecha del tiempo" in *Anales de Geografía* de la Universidad Complutense, Madrid (2007; vol. 27, pp. 29–44).

The current studies in Barcelona, a place that is becoming a relevant immigrant pole during the last 15 years for many young people coming from Brazil and the rest of Latin America and the Eastern Europe, allows us to know some of the features of the end of migratory stories. The interpretation of both aspects, from the origins and from the destination, has been encouraged by the preoccupation that has arisen around the events of French *banlieues*[6] during the autumn of 2005, where many young people belonging to the second and third generation of immigrants and feeling marginalized, reacted with despair in a context of non escapist possibilities (Tuan 1998).

The first question analyzed is why an increasing number of young people, with absolutely different socio-economic and cultural profiles, make the same decision to emigrate to the United States or to some countries of the European Union. One response is a generalised lack of local or national conventional well-being. We try to demonstrate here the hypothesis that there also exists a generational component, due to the unique senses of place developed by young people and a general feeling of social exclusion. One motivation to this theme probably could be found in our own preoccupation for the evil functioning of our universities, which tend to permanently exclude the best young people formed by us, sending them to an undefined market of nonqualified tertiary jobs.

The second question is with respect to the fact that the concept of youth has dramatically changed in the last few decades, essentially because the increasing difficulties and chronological delay of the integration of the young people into the regular labour market. This general delay, in addition to the problems of structural unemployment that it implies, has created some kind of enlargement of the adolescent period for the majority of individuals. The traditional demographic age classification for paid labour, with its limits at 15 and 65 years, has been completely over passed. New university study opportunities reflect the rise of the international exchanges, masters and PhD degrees everywhere (not only in the Anglo-Saxon world where it was already traditional); this, together with the increase of many different informal and temporary jobs allow many contemporary people to maintain the typical *irresponsibility* of youth after 25, even 30 years if no more.[7] This fact has permitted the coexistence of generations of young people in the same situation, creating both new ties of solidarity and competition among them and a clear rupture with the world of adults in general.

The Bosnian Exodus, A Very Special Context

The Bosnian writer Dževad Karahasan put the word *exodus*, with both biblical and Zionist echoes, into the title of his impressive book on the siege of Sarajevo from 1992 till 1995 (Karahasan 1994). He tries to explain the unexplainable fact of the recent Balkan war, with its renewed ethnic cleansing after a long period of a relative

6 *Banlieu* is the French name for the poorest and peripheric suburbs of the big cities, mainly built on the principles of the Modernist planning.

7 It is the known complex of Peter Pan, the character created in 1904 by the Scottish writer J.M. Barrie (1860–1937).

constructive conviviality, under the Yugoslavian Tito's regime. Karahasan, born in 1953, could truly represent the intermediate age Bosnian generation[8] that is looking for a mythified past in order to find a solid base for a certain future of hope. The oldest generations, on the contrary, were more conformist with an almost traditional misfortune, and seem to be able to endure all kind of situations. But the young Bosnian people exhibit another attitude, absolutely different from both ancestral generations.

The dominant trait in the attitudes of the young is the negativity felt with regards to all kinds of things related with ideas or habits of their ancestors. In spite of the historically contradictory evolution of modernization in Bosnia, with brief but progressive and positive periods (Carreras and Moreno 2007), young people don't want to give any credibility to the generations that, in fact, originated the recent Balkan conflict (1992–95). They publicly affirm their desire to forget all their past and to enjoy today's everyday life even with its scarce possibilities, making real the sentence that the past needs to be a foreign country.[9]

The methodological and even ideological difficulties inherent in the discussion of young people's attitudes are many. We undertake here an analysis of the everyday life of different groups of young Bosnians, mainly in post-war Sarajevo where urban life has been reconstructed – mainly with the support and control of the members of a very strong foreign army intervention. In this direction, we have explored the reconstruction or rebuilding of different senses of place in Sarajevo after the war, mainly focused from the images of the city and from the mental maps of young people.

Both for good and evil, Sarajevo has succeeded to develop and to diffuse its own images into both the national and international collective imagination. In spite of the current popular geographical ignorance, Sarajevo has been put on the map,[10] at least since the very beginning of the Twentieth Century. The different images of the city could be organized in two different and separate parts: a) the traditional ones (from 1878 to 1992), and; b) the new ones (since 1992).

The Traditional Images of Sarajevo

The first general and traditional image that must be taken into consideration is the great attractiveness of the place itself as a general and beautiful landscape.[11] The urban site of Sarajevo is in many aspects unique. The location of the city at the bottom of the Miljacka valley, and its historical growth along the river has created an image of morphological and historical continuity and linearity that permits the

8 Just *my* generation.

9 This sentence comes from Leslie Poles Hartley (1895–1972) and constitutes the title of an important book by David Lowenthal (Lowenthal 1985).

10 During the years of the organization of the Summer Olympic Games of 1992, the city of Barcelona had ironically invested a lot of money and cultural capital in order to be put on the map of the world's collective imagination.

11 I used to say that it is very difficult to visit the city only one time; we were coming back year after year, even the terrible September of 2001. My students are also periodically applying for grants to do fieldwork there.

reconstruction of the history of the city by calmly walking, strolling along its streets. The surrounding mountains appear as a protective nest for the city and, by night, the lights of the suburban houses design their own Sarajevian sky, with a myriad of artificial constellations. This attractiveness plays an important role in facilitating the development of strong senses of place among the inhabitants and visitors of the city.

A second and essential image is that of its European Imperial provincial past, expressed through the most symbolic buildings. The Austro-Hungarian rule, from 1878, gave to the traditional Ottoman city not only a new and impressive architecture but also a totally new way of life. Austria implemented the new capitalist system in its many forms, including: nuclear families, social segregation, apartment buildings, bourgeois shops, monumental buildings, and public spaces (streets, squares and promenades). It is very interesting to point out that some Austrian architects have even created a very romantic version of Muslim buildings.

A third often more manipulated image is a spiritual one, that being the European Jerusalem; this image is much diffused through books and maps (Kostovic 2001). The togetherness of different Christian religions (mainly western Catholics and Slavic Orthodox) with Islam and Judaism, at least till the beginning of the Second World War, helped to develop an image of peaceful and, in many ways, mythical multiculturalism, which Karahasan also maintains in the background of his previously cited novel.

A fourth, and a very different international image, comes from the dramatic and punctual incident that symbolically started the First World War, very close to the historical and central Latin bridge that still generates bad feelings among the local population today. In this regard, it is very interesting to consider the periodical changes in the evaluation of the actions of the Serbian nationalist Gabrilo Princip, sometimes considered a national hero, sometimes almost forgotten (like nowadays). A similar and symptomatic change in public consideration could be also found around the works of the Serbo-Croatian Bosnian winner of the 1961 Nobel Prize in Literature – Ivo Andrič (1892–1975), who was a Bosnian and depicted magnificently the traditional Bosnian life in some of his most famous books (Carreras 2003). On the contrary, it must be underlined that the unique image always present in one of the most important and central streets of Sarajevo, the Marshala Tita Avenue, is dedicated to the former leader of socialist Yugoslavia.

The preservation of its name could reflect the awareness of the international prestige of the country during the socialist era and especially of its international role in leading the Non-Aligned Movement.[12] The socialist period is generally connected to grey and old-fashioned images, but the older generations know the relative welfare and freedom of the Titoist regime against the Soviet model, creating contradictory feelings and images.

Finally, and as a sort of culmination of the Yugoslavian period, the image of the 1984 Winter Olympic Games and its remains always plays a very important role on the city and its surroundings. A clear transversal axis of modern buildings marks the

---

12 The Non-Aligned Movement was created at the international conference of Colombo (Sri Lanka) in 1954 assembling the initiatives of India, Ghana, Egypt, Indonesia and Yugoslavia. Marshall Tito organized the 1961 second summit in Belgrade.

new centre of the city, connecting the historical and the post World War II extensions. This fact helps to also explain the general[13] sympathy of the city of Barcelona and its Olympic mythification to Sarajevo and even the Catalan investments in the first reconstruction period of the city.

The New Images of Sarajevo

The Yugoslavian war, specially the Serbian siege of Sarajevo (1992–95), has offered to the whole world terrible and different images that up until today informs literature, cinema and human suffering of all sorts. The mythical nest of mountains of Sarajevo has become, in this occasion, a criminal trap. The 1993 book of Miroslav Prstojevic has diffused many of these images of brutality. War always remains related to Sarajevo's present image, even generating some kind of philanthropic international tourism.[14]

New images are related to the prevalence of ethnicity over other social and cultural values, like family, residence or job. Bosnian is the only community of the former Yugoslavia that does not have support of other states in the area, like the Croatians and the Serbians, and Bosnians are therefore obliged to develop strong internal images of identity. This creation becomes more difficult considering the present theoretical background of the crash of civilisations (Huntington 1996) that conflict against Islamic people, in general. The international military intervention (NATO, OSCE) complicates an already complicated scene.

One of the most visible elements of this image of contemporary ethnic conflict is that of the reconstructed churches which have been financed separately by the different states on a clear cultural basis. Greece financed the reconstruction of the Orthodox Cathedral, Saudi Arabia and other Arabian countries financed the numerous mosques, characterized with totally different shapes from the Bosnian originals (there is even one designed with a Malaysian shape). After the first period of this kind of elemental rebuilding, a more general reconstruction has started; in addition to demolishing a number of buildings ruined by the war, a number of buildings have been renovated, including the Unitič centre, Avaz.[15] Only the former Austrian local administration building, which later became the Bosnian National Library, remains

13 As usual, after some years the solidarity has been diverted to other international conflicts (Kossovo, Palestine, etc …).

14 On the basis of many solidarity non-governmental organizations, it could be sometimes found a new kind of tourism very popular among young western students. We found a very similar kind of tourism in present Cuba, but with opposed purposes. Our colleague and co-author George Ashworth has published a very interesting book on other new and peculiar tourist attractions (Ashworth and Hartman 2005).

15 This Croatian business and commercial company has completely renovated the original building of the Bosnian journal Oslobodenje (freedom), that has been published every day all the siege along.

today as a war memorial in spite of the publicity of the financial help of the Austrian government for its reconstruction.[16]

So in less than ten years the physical reconstruction has been almost completed, faster than anybody would have thought; this has offered new images of the city. The role of the new foreign embassies has also helped to partially restore the traditional Austro-Hungarian image of an imperial provincial capital. A constellation of alive and noisy mosques contrasts the image of a European city with an oriental touch. Finally, the foreign investments reinforce the diffusion of global capitalism, which is very visible in the new peripheral shopping and business centres.[17] Altogether, the City feels as if it is a sort of a coming back to the beginning of the 20th Century, which is when the first diffusion of Capitalism happened in the country (Carreras and Moreno 2007),[18] producing the traditional image of Sarajevo as an European provincial capital.

At the same time, the presence of a big foreign army provided another set of new images, both directly – with the high visibility of military uniforms and cars with different national flags, and indirectly – through the economic tourism activities reflecting the army's great number and western consumption demands.

Finally, the lost occasion of the organization of the winter Olympic Games for 2010 has been one of the international failed images of this new era. The reconstruction of the Olympic infrastructure and the development of a regional winter tourist centre has not been reason enough to convince the International Olympic Committee, whom chose Vancouver. Sarajevo confronts the competition among cities on the international arena – many of whom been strongly developed under a global economy, with the clear handicap of political instability.

Other elements could create other images that otherwise become very common in many other contemporary countries, and are relatively unspecific to Sarajevo or to Bosnia and Herzegovina. Given this direction, it must be specifically pointed out that the general trend among young people is to put themselves out of the system via their generic interest in economic or political foreign emigration, as a form of escapism (Tuan 1998).

Nevertheless, the final balance of the different sets of Sarajevian images most probably reinforces a negative image for foreigners, from Sarajevo assassination (1914) to the Serbian siege (1995). Among the locals, nobody wants to remember the past centuries, not even the Bosnians themselves. Locals are not aware of the new landscapes of post-socialist and post-modern capitalism, with its provincial

---

16 In the last springtime days, the Spanish journal *El Pais* published that the Spanish government was looking for financing the final reconstruction of this last memorial monument.

17 The new shopping centres are symbols of globalization and play an important role in the diffussion of Western images that are changing the consumption values of Sarajevian people, which is starting to be analyzed (White and Absher 2007)

18 In this article the authors try to demonstrate the hypothesis that modernization is a concept more complex than economics alone; Islamization in the 15th Century could had been the initial form of modernization for the country with its integration to a large and new commercial economy – that being the Ottoman Empire.

features financed by foreign countries, many of which are the closest neighbours and ironically ancient partners of the former Yugoslavia.

The Individual Images: The Mental Maps

Mental maps analysis is a relatively ancient and useful technique among the social sciences,[19] used as a way of knowing about the feelings (such as sense of place) that people have with relation to their places. The main difficulty use to be the general poor skill of designing maps among ordinary people; for this reason, the research has been centred in the maps created in our summer university courses, with the help of some of its teachers.[20] A first classification of the material differentiates 139 maps,[21] those created by foreign people (39) and those created by Sarajevian and Bosnian people.

The external vision of the students who were foreigners in Sarajevo (mainly Spanish in this case) is not very interesting, given the scope of this research, but allows us to do a first comparison and to confirm the importance of the international images of Sarajevo. They are used to represent the city at a very small scale, reflecting their limited general knowledge, and are most representative of common tourist

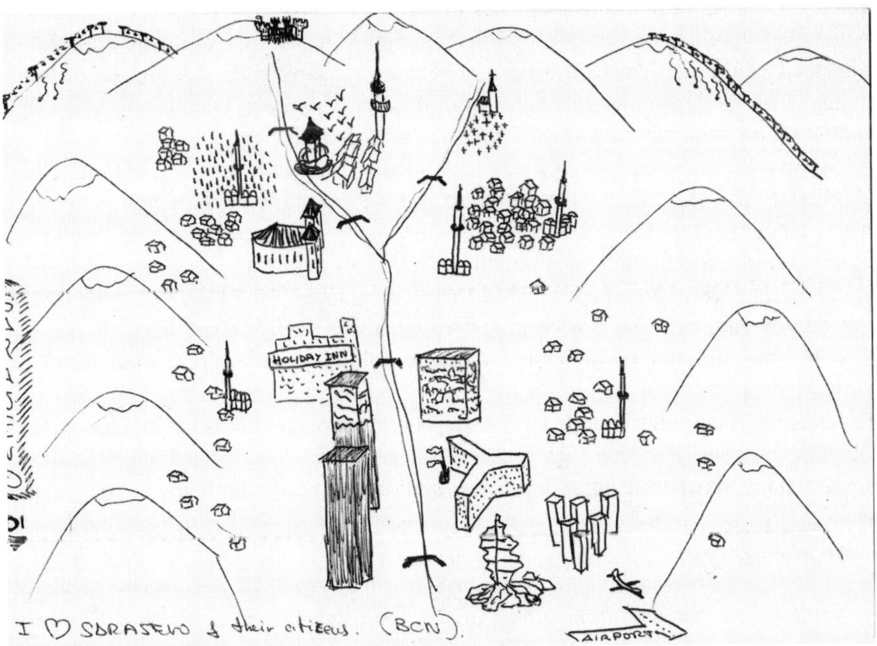

Figure 9.1 Foreign Sarajevo mental map

19 The most interesting approach is Gould and White, 1974.

20 The author must to thank specially his colleagues Elisabet Rosa and Eva Serra, teachers and professionals in G.I.S. for their collaboration in this project.

21 The total number of mental maps equalled almost 300, but many of them were not useful because they were either too similar or illegible.

sites, such as symbolised by the omnipresence of the mosques and the war memories between the snowy mountains. When in the plane landing in the international airport, many observed very important local features, such as the course of *the* river; even the orientation of the whole map to the east reflects a foreigner's use of the City plan, having no previous City experience and knowledge. At the same time, it must be underlined that these mental maps are also a form of representation of the already discussed attractiveness of Sarajevo, clearly reinforced by the sentence that someone wrote below their map: *I love Sarajevo and their citizen*s (in spite that probably the author loved only a special one of them …).

The internal visions, on the contrary, are normally more complex. The common orientation of Sarajevo mental maps is alike the usual maps of the City, with the north on the top; this is likely due to clear influence of scholar instruction. But even when the south appears on the top (like in old cartographical Islamic tradition), the river always runs parallel to the base of the sheet, clearly indicating the linearity of the city and the fundamental role of the Miljacka (very often with its name) in the city's communications system and in the orientation process of everyday behaviour.

The complexity and richness of information, in general, has created many difficulties in the analysis of the mental maps of Bosnian students, obliging us to discuss some aspects with the authors in a few collective meetings (one per year). A detailed analysis of the one hundred mental maps allows us to determine the relevance of the predominant images. The main reason for the dominance of these images may be the challenging issue of mobility in this long linear city, particularly for the young who have scarce access to the use of private cars.

From a geographical point of view, it has been possible to do a first and clear dual classification of the Sarajevian mental maps. One group of young people focus their maps exclusively on the historical centre of the city (almost 70 per cent), while the other focuses only in their own neighbourhood (30 per cent). This peculiar feature could be representative of a relatively fragmented social life of the young people in this City during this first post-war period. However, this fragmentation could also be coherent with some consequences of the traditional socialist urban planning. This planning was based on the building of quite isolated neighbour units, complete with their own services and facilities, including the omnipresent mosque and associated frequent prayers. Because this fragmentation and its socialist origin, the author Karahasan dislikes these neighbourhoods and does not consider them as truly urban, the latter which would only be the central historic part of the city (Ottoman and Austro-Hungarian).

This next mental map could represent the first group of City representations that only reproduce the historical centre of the City, that being the famous Basarčija (despite the fact that this is a commercial street). The central square appears with a figurative fountain design and the omnipresence of pigeons – a very common urban animal and, at the same time, a more or less vague symbol of peace. The river is always there with its name clearly marked, almost supporting or integrating the different elements of the map. Some very significant monuments underline traditional Sarajevian multiculturalism: the Catholic Cathedral, a mosque, and the Bosnian National Library, the latter which was destroyed and not yet reconstructed. Finally, two different cheap restaurants clearly mark the main use of this space by

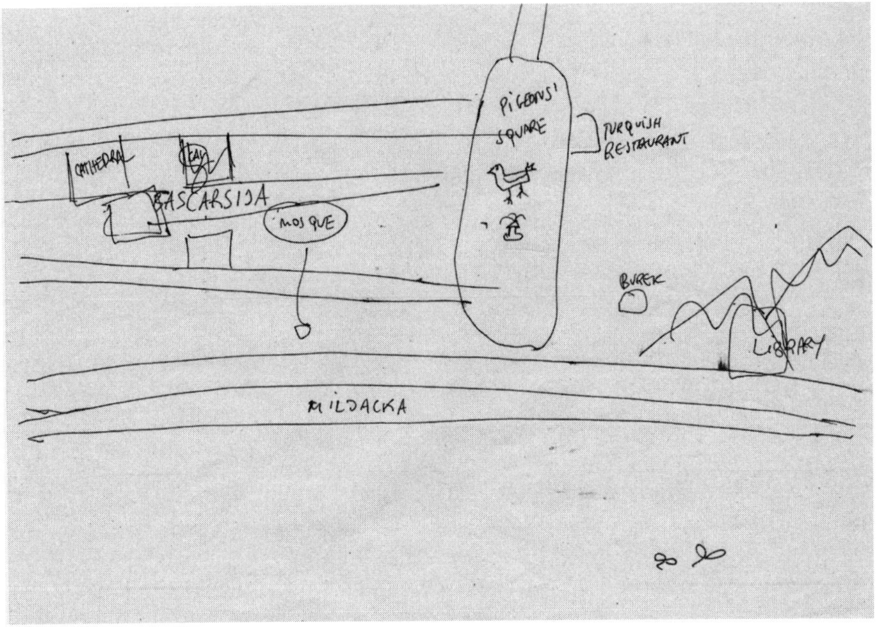

Figure 9.2 Mental map of Sarajevo's old centre

young people: a Turkish establishment and a bureč, which is traditional and sells inexpensive Bosnian snacks.

The smaller group of mental maps presents the relevance of the young people's neighbourhood life, centred on the Alyspasino polje quarter which is situated on the western part of the new city of Sarajevo[22] (Figure 9.3). The river Miljacka is always there, in the centre of the map, underlined by a dot frame and with its name written two times. The main road, to the centre (not indicated) and to Cengic runs parallel to the river, centring all the elements represented. The designed urban area here is even smaller than that of the previous central map, but the number of details is considerably larger. This fact could indicate the quantity of time spent in their own quarter, which is not contradictory with frequent but rapid visits to the centre for leisure purposes. The details involve buildings, place's names, services (Christian church, emergency post, tram stations) and even include the location of the home of the mapper, as well as the location of the Transport Faculty where the study was carried out. Some bars and the Zegin stadium complete the mental map, signed by the author, despite recommended anonymity.

These two different typologies of mental maps clearly indicate different use and appropriation of the city spaces by different individuals while, at the same time, reflects common features.

---

22 The neighbourhood of Grbavica is also located (and indicated) on the map. This neighbourhood has been played in the film of Jasmila Zbanic (2006) which reflects similar aspects of the everyday life of people in contemporary Sarajevo, as well as the conflicts among generations.

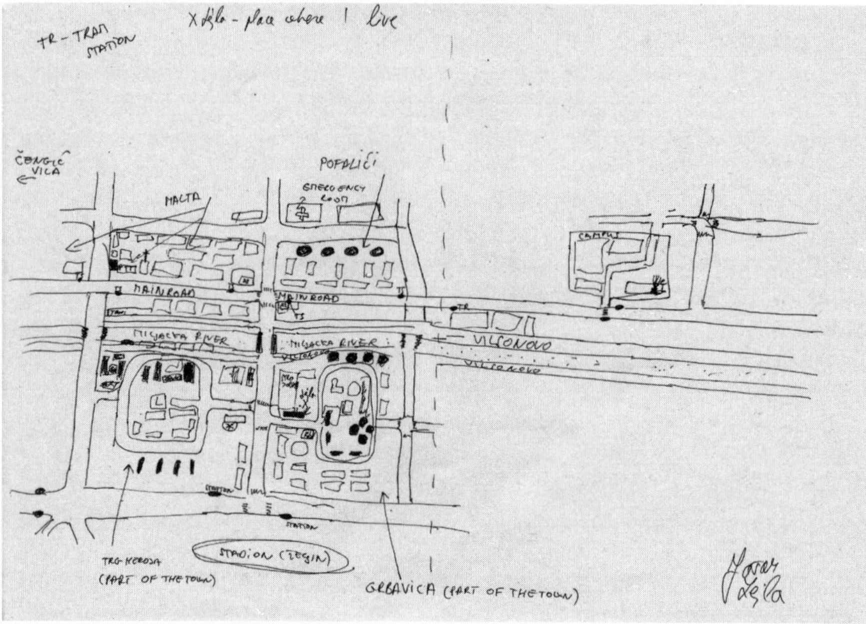

Figure 9.3 Mental map of Sarajevo's neighbourhood

First of all, public space is predominant while the domestic spaces are only designed as black boxes. This is a consequence of the questions regarding public life and of a deep sense of privacy, while suggesting a new area of research particular to the material conditions of the domestic everyday life of Sarajevian people. From this point of view, Sarajevo seems to be really beloved by all the participants, reinforcing the idea of the fascination of the City discussed earlier and feeding any kind of sense of place. In this context, young Sarajevian people also seem to be relatively insensible, enjoying their social life and only looking for amusement through walking and going to bars and discos. This next map, which is a very simple a linear one (a bus itinerary mainly), puts the bar at same level that all the other milestones.

When speaking with the participants, the majority of the young people seem to be dreaming about migrating to the European Union or, preferably the United States. They try to share the general welfare of these countries, as reflected in their collective imagination, built from images captured on TV and cinema. These dreams could also be read from the permanent occupation of the central Sarajevo streets and squares; many people are found walking and strolling all day and night in spite of the weather conditions, which are very severe during the long winters. Although appearing like a typical consumerist way of life, there doesn't exist the required material conditions.

Directly connected with this is the importance accorded by young people to love, in general. Hearts and other love symbology appears everywhere in the maps, indicating the search for pleasure and joy, possibly accentuated among the Sarajevian young. In the same way, different animal designs give a vital impression to the maps, such as pigeons on the Main Square or fish in the river. Discussing these aspects with the students resulted in responses that spoke of the need for individual affirmation and disregard for collective solutions. This is an example of the society of individuals defined by the German philosopher Norbert Elias in 1987, which is not so different from other cases like Sao Paulo, Napoli or even Barcelona (Carreras 2003; 2006). At the same time, the noble sprit of the young appears on the non-required signature of almost all the maps. This fact implies the apparition of some kind of generational trend to migrate in search of pleasure that may not be attainable via local approaches. At the same time, this gives an impression of the social life in Sarajevo that relates to the atmosphere of Belle époque, typical of the postbellum periods. A clear and strong sense of place among young people is not able to break the interest in emigration.

A second important fact emerging from these mental map analyses is related to the continuity of the traditional cultural fragmentation of the City. In spite of many formal affirmations of a new kind of multiculturalism that gives all the responsibility of the war to the oldest generations, little detail of this is provided in the maps. Further, public debates demonstrate the difficulty of speaking about the country and its recent history. In this regard, it was possible to see propaganda against the works of Den Haag International tribunal for former Yugoslavia in certain periods but, at the same time, was confirmed everywhere via the use of the traditional symbols and the general ignorance of the new federal flag, significantly named the Westendorp flag.[23]

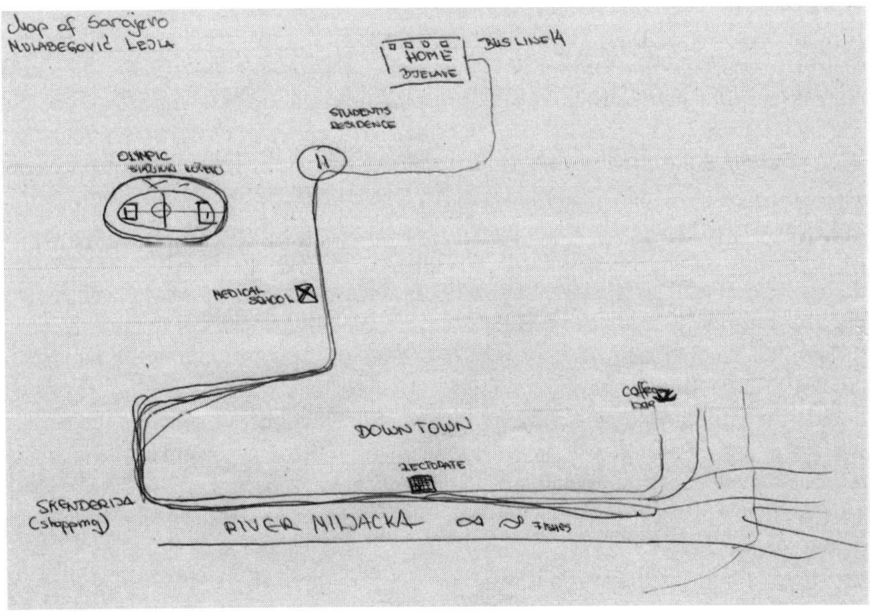

Figure 9.4 Linear mental map of Sarajevo

23 Named after its creator, the UNO High Commissioner for Bosnia and Herzegovina in 1997–99 – the Spanish ambassador Carlos Westendorp.

Conclusions

The results of this research underline the relevance of sense of place in achieving a minimum quality of life and well-being. By easily identifying sense of place within their own city and neighbourhood, the young seem to be able to counterbalance many material difficulties of their everyday life. At the same time, the psychological footedness that implies sense of place doesn't succeed in stopping the escapist solution – the decision to migrate; sense of place exists as a dualism between western dreams and Bosnian everyday life. The good health that characterizes youth provides confidence in migration, with their immortality minimizing any negative elements.

Parallel and in some way connected to the analysis of senses of place, is the significance of a special and differential *sense of time*. The expansive heritage of Islam or the influence of the Mediterranean culture characteristic of the Neretva Valley, helps to explain the special experience of time for the Bosnian people; even among the young. Time is running visibly less fast that in the majority of western societies. The five calls to prayer from the four corners of the city imply a time measurement absolutely different from the mechanical and exact clocks used elsewhere. This calm or slowness probably helps to explain the habit to walk even in the cold winter nights in the Sarajevian public spaces. Time appears embodied in the space and sense of place is not apart from it (Massey 2005).

An important theoretical question arises from the conflict between sense of place of the old and traditional residents in Sarajevo, and that of the new residents – defined either by age or by recent immigration. According to the 1995 Dayton Agreements, the continuity of what have become Sarajevo's collective traditional images would be desirable, but the new social situation and the wants of the local population offer many difficulties to any kind of continuity. Nevertheless, the old and the new residents are able to develop their own senses of place, throughout their appropriation of the city or some parts of it – with many different meanings, rhythms and contradictions, and according their social, ethnical or political characteristics.

A methodological question also arises from this research, that being the usefulness of the collective images and the mental map analysis in understanding the perceptions of people and how they explain variation in senses of place and the decisions to migrate or to travel. The difference in the scale of images between foreign and local citizens suggests the interest to promote medium scale images of the city (Sarajevo as a totality) among local inhabitants. This may be implemented in order to help to develop other complementary and collective senses of place that could play a significative role in their attachment to the country and act as a barrier to escapism. At the same time, general and vague images of western welfare become stronger than any local image or sense of place, reinforcing the decision to migrate in many countries.

There is one other important conclusion with regard to the general feelings of young people during this period of globalization. Coming from the securities of the 1960s to the uncertainties of the 1990s, from the end of the Fordist system to high modernity, many changes have been experienced. In this historical context, the desire to migrate becomes not conjunctional, but structural. This fact implies that the sense of place not could be based on job or life stability anymore, but mainly based on the place and its qualities and attributes (specially the sense of time related to these spaces). Thus sense of place appears to be more and more strategic

Finally, the hypothesis expressed at the beginning of the chapter with regard to the evident general generational break, is confirmed. It may appear exaggerated in Bosnia because the tragical events of the pre-war period and the war itself. Nevertheless, we must conclude that in our contemporary society of individuals and consumers, many values that characterize youth have been stolen by the older generations. This could be proven in the generalization of the body culture and medical, aesthetic care among adults and older people, or in the reluctance that adults have to leave jobs. In the same way, the integration of apparently endless technological progress allows the majority of the old generation to assume all kind of innovations that traditionally were a clear mark of youth.[24]

Fragmentation appears as the new social trend. Fragmentation of spaces reinforces the large scale and the deteriorated socialist neighbourhoods; this includes the fragmentation of cultural groups (traditional and new orthodox Islamist; Islamists and Christians; Catholics and Orthodoxies; Bosnians and Serbs and Croats) and the fragmentation of age groups.

It is also possible to make a more positive conclusion. In this regard, it is necessary to suggest different ways in which many aspects of sense of placc can bc implemented and spread at the local level, impacting regional, national and international policies. In this way, it is very important to recognize the significant impact of the production and consumption of collective images that could potentially define these policies and, if implemented, potentially reconcile people with its spaces across generations.

References

Aguilar, C. and Molina, J.L. (2004), 'Identidad étnica y redes personales entre jóvenes de Sarajevo', *Araucaria (Buenos Aires)* 6(12).

Ashworth, G.J. and Hartmann, R. (2005), *Horror and Human Tragedy Revisited: The Management of Sites of Atrocities for Tourist* (New York: Cognizant).

Bublin, M. (2004), *Rehabilitacija gradova Bosne i Hercegovine Sarajevo 2003* (Sarajevo: Buybook)

—— (2005), *Gradovi Bosne i Hercegovine. Milenijrazvoja i godine urbicida (Bosnian-English second edition)* (Sarajevo: Sarajevo Publishing).

Carlson, M. and Listhaug, O. (2007), 'Citizens' perceptions of Human Rights practices', *Journal of Peace Research* 44: 465–483.

Carreras, C. (2003), *La Barcelona literària* (Barcelona: Edicions Proa).

—— (2004), 'Sarajevo, ciutat màrtir', in C. Carreras (ed.), *Atles de la Diversitat* (Barcelona, Enciclopèdia Catalana)

—— (2006), 'São Paulo una sola ciudad? Las identidades fragmentarias de la metrópolis globalizada', in C. Carreras and A.F.A. Carlos (eds), *Barcelona y São Paulo cara a cara* (Barcelona: Ed. Davinci).

24 That becomes clear even at the university world, where old professors remain active for years, while the young researchers are excluded almost definitively generation after generation.

Carreras, C. and Moreno, S. (2007), 'Los procesos de modernización en Sarajevo. La incierta dirección de la flecha del tiempo', *Anales de Geografía* (Madrid: Universidad Complutense) (in press).

Crnobrnja, M. (1996), *The Yugoslav Drama* 2nd Edn. (London: I.B.Tauris Publishers).

DD.AA.(117) *Sarajevo: photo-monograph*, IP Svjetlost, Sarajevo.

Dell'Agnese, E. (2002), *Geopolitica dei Balcani: luoghi, narrazioni, percorsi* (Milano: Unicopli).

Elias, N. (1987), *Die Gessellschaft der Individuen* (Frankfurt: Verlag).

Eyles, J. (1985), *Senses of Place* (Warrington: Silverbrook Press).

Gould, P. and White, R. (1974), *Mental Maps* (London: Penguin Books).

Huntington, S.P. (1996), *The Clash of Civilizations and the Remaking of World Order* (New York: Simon & Schuster).

Karahasan, D. (1994), *Sarajevo, Exodus of a City* (New York: Kodansha International).

Kostovic, N. (2001), *Sarajevo. Europski Jeruzalem* (text in Bosnian, English and German) (Sarajevo: Bravo Public Team).

Lowenthal, D. (1985), *The Past is a Foreign Country* (Cambridge: Cambridge University Press).

Massey, D. (2005), *For Space* (London: SAGE Publications Ltd.).

Mavric, S. (2000), *Mom, What is War?* (Sarajevo: Buybook).

Moreno, S. (2007), 'Ferhadija, la calle elegante de Sarajevo', in S.M.M. Pacheco and C. Carreras (eds), *As multiplas abordagems da rua comercial* (Rio de Janeiro: UFRJ) (in press).

Petonnet, C. (1982), 'L'observation flottante. L'exemple d'un cimitière parisien', *L'Homme* 22 (4): 37–47.

Prstojevic, M. (1993), *Survival Guide Sarajevo* (Croatia: FAMA).

Tuan, Y.-F. (1998), *Escapism* (Baltimore: Johns Hopkins University Press).

White, D.W. and Absher, K. (2007), 'Positioning of retail stores in Central and Eastern European accession states: Standardization versus adaptation', *European Journal of Marketing* 41: 292–306.

Wilmer, F. (2002), *The Social Construction of Man, the State and War: Identity, Conflict and Violence in Former Yugoslavia* (New York: Routledge).

Chapter 10

Sense of Place and Quality of Life in Post-socialist Societies

Marko Krevs

Post-socialist societies have experienced quite fundamental economical, social, and environmental transformations since 1980s.[1] Their specific geography and history, and the specificity of their transition from socialistic towards capitalistic and democratic regulation are reflected in some distinct characteristics and changes of quality of life, health, and senses of places.[2] But it is almost impossible to think of post-socialist societies as a homogeneous community. Instead of searching for what they have in common, extensive review of the literature (published separately, Krevs 2008) has been carried out to collect a mosaic of selected examples highlighting some national, regional, or local specificities within the studied context. In this chapter two of the studies are exposed in some detail, revealing relations between sense of neighbourhoods and quality of life in urban municipality of Ljubljana, Slovenia, and between sense of 'Mediterraneity' and health in Slovenia. Other selected examples are used in the introductory overview, and as a context in the discussion about these two local examples within a wider post-socialist frame.

Many languages and some non-Latin alphabets are used in post-socialist countries, and only a part of the relevant research results about these countries may be expected to be accessible. Besides, quality of life, health and sense of place are studied in a number of disciplines, using various concepts and methodologies.[3] Consequently

1 Term post-socialist societies generally refers to the countries in the process of formal transformation from the former socialistic regulation. Term is usually used to address the post-socialist countries of Central and Eastern Europe, and those that have evolved from the former Soviet Union (Commonwealth of Independent States, CIS). Former Eastern Germany is sometimes 'missing' from the list because of its merger with highly developed Western Germany, consequently loosing the post-socialist country identity. Nicaragua could fit, being defined as socialist country in the Sandinistas period between 1979 and 1990. But it is rarely put into this frame, probably due to its geographic remoteness. On the other hand, China which officially still holds the socialistic regulation, allows substantial economic and social changes that are far more extensive than in many post-socialist countries.

2 Term communistic is sometimes used instead of socialistic. In Marxist theory socialism is the intermediate state in transition between capitalism and communism. In that sense all the studied countries have been socialistic, therefore term post-socialist countries is used in the chapter.

3 Professional groups like 'Post-Socialist Geographies Research Group' and 'Soyuz: The Research Network for Postsocialist Cultural Studies' play an important interdisciplinary role

very loose definitions of the three terms had to be applied in the construction of the above mentioned mosaic of examples.

Quality of life directly or indirectly represents a state of social wellbeing identified by quantitative or qualitative information, reflecting the circumstances of living and their perception. Usually it is represented in a complex way, with many versions available on a national, regional (e.g. welfare, level of living, Human Development Index), or personal level (e.g. happiness, or rehabilitation of an injured person). Health may be represented by physical or mental (ill-)health status indicators, and indirectly through the assessment of health practices or coping skills, or the availability of health services. Sense of place is usually expressed by positive or negative sentiments in relation to specific places, but also as a character intrinsic to a place as a localized, bounded and material entity (McDowell 1997). Some of the terms used to reflect sense of place include: geographical, spatial, regional or local perception or identity, perception of place, attachment to place, topophilia, and topophobia.

At least partly relevant works have been found for all post-socialist countries. They have been arranged into three categories. The first contains the works related mainly to quality of life and health, the second mainly to senses of places, and the third to works demonstrating more evident linking of the two themes. This last category – the focus of this book, is relatively poorly represented from the point of view of the number of the studies available, as well as due to the concentration of their focus on the western part of 'the post-socialist region'. A brief overview of the first two categories of studies is therefore necessary to provide a better coverage of the breadth of the region, as well as some details specific to the quality of life, health, and sense of place issues in the post-socialist societies.

Quality of Life and Health in the Post-socialist Societies

At the beginning of the transition period, the post-socialist countries – except the Central Asian newly independents states, fared quite well in many ways, including their: demographic profile, employment, education, and health. Throughout the 1990s, the effectiveness of the state drastically deteriorated in many countries, evident in: gross domestic product falling up to 70 per cent, growing income inequality, full employment ceasing to exist (Fajth 2000, 76–77), reducing life expectancy, and the emergence of poverty as a major issue in the region (United Nations Economic Commission for Europe 2004).

Since 2000, all the post-socialist countries have returned to positive economic growth, but significant improvements in material and social welfare 'of ordinary people' remains elusive, since 'this ultimate objective of transition … appears to have taken second place to the imperative of economic growth' (ibid 163). Nevertheless, the core socio-economic factors are found to have important impact on happiness (Hayo 2007) or life satisfaction (Fahey and Smyth 2004) in post-socialist

in bringing together the researchers, focusing on post-socialist countries and also on topics addressed in this book.

countries who have experienced these dramatic changes. In addition, our analysis of the correlation between 'objective aspects' of quality of life – measured by the Human Development Index (UNDP 2006), and economic conditions – measured by gross domestic product at purchasing power parity per capita, appears to be much stronger for post-socialist countries ($r=0.94$, $p<0.001$) than for all countries listed in the Human Development Index database ($r=0.75$).[4]

To put 'objective quality of life' in post-socialist countries into the frame of all World countries, the Human Development Index (HDI) may be used.[5] Post-socialist countries are quite wide-spread across the HDI scale: about 15 per cent of all the countries are better ranked than Slovenia, and 69 per cent are better ranked than Tajikistan, representing the two extreme examples.

The differences among the post-socialist countries are illustrated on Figure 10.1. Gross domestic product at purchasing power parity per capita presents the material aspects, and life expectancy at birth portrays the health dimension of this simplified view of quality of life (UNDP 2006). The correlation between these two aspects of 'objective quality of life' is considerably lower than between the economic situation and the Human Development Index discussed above. This result may be complemented with the findings that the satisfaction gap between the better-off and poorer post-socialist countries, included in Quality of life in European research, is the largest for the health care system (Alber et al. 2004, 83).

The countries in the upper right quarter of the Figure 10.1 are expected to be dealing with several quality of life and health problems/challenges quite similar to those in the economically developed countries. In other countries, the specificities of the post-socialist transformation seem to have much stronger effects on quality of life. In all the countries with a very unfavourable economic situation, there is an expected poorer ability to cope with quality of life and health related problems.

The findings of a few complex studies of objective and subjective quality of life in post-socialist countries have already been discussed, but some of the post-socialist specificities are especially noticeable in the studies that focus on narrower aspects of quality of life or health. The socially undesired – and for many individuals also tragic, issues prevail in the reviewed literature (Krevs 2008); the positive aspects of the post-socialist transformations obviously can not compete for the attention of both researchers and the ordinary people. Populations like the elderly, the unemployed, and former refugee children are especially vulnerable, and consequently often the focus of studies of the effects of post-socialist transformations. The effects of economic and war-related migrations on quality of life range from brain drain to human trafficking and sex slaves. Gender related studies emphasize problems of female employment, informal work, domestic violence, and discontent with life as

4 In the Human Development Index (UNDP 2006) database 177 countries are assessed.

5 Several other sources of complex measures of quality of life are available today, taking at least some of the post-socialist countries into account, like European Quality of life survey (Alber et al. 2004), Sustainable human development index (Costantini, Monni 2005), indicators of subjective well-being for European countries, collected in 1999–2000 European Values Study (Fahe and Smyth 2004), Quality of Life Index by International Living (2007), The Economist Unit's Quality of life index (EIU 2005), and Happy Planet Index (NEF 2006).

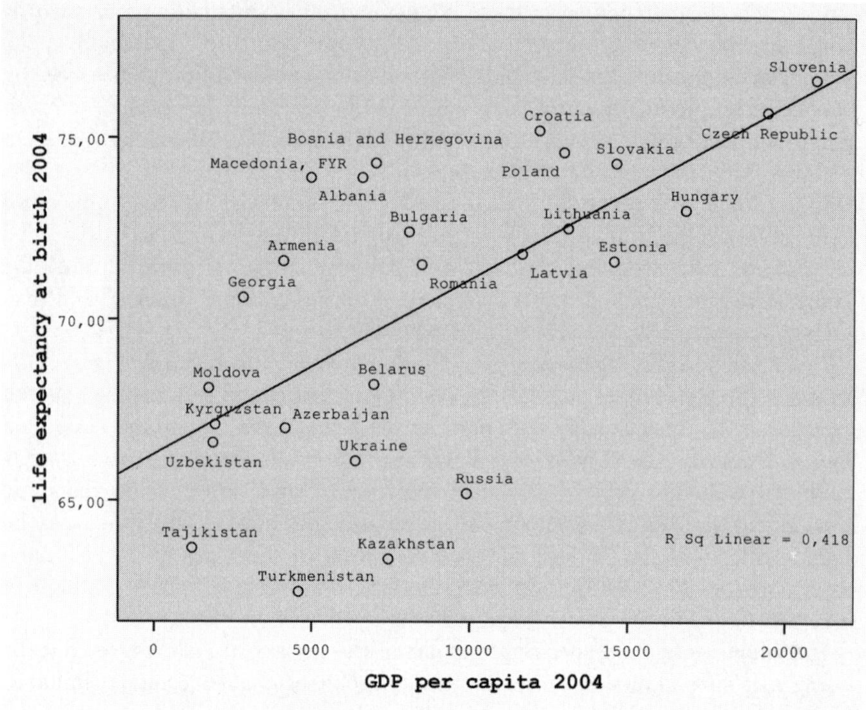

Figure 10.1 Gross domestic product per capita and life expectancy at birth in 2004 by post-socialist countries

Source: UNDP (2006).

factors of declining birthrates. The effects of sexual violence – used as a means of genocide, are studied specific to the 1990s wars in former Yugoslavia. Specific relations between urban and rural areas, and their changes within the transitional processes, are studied from the perspective of quality of life in many of the post-socialist countries. In the search for particular aspects of the post-socialist transition, nostalgia for 'the communistic times' should not be overlooked, expressed not only as a feeling, but as a driving power of trade with communistic era souvenirs, music, mobile phone ringtones, and supporting socialist parties back into power.

Post-socialist transitional processes may drastically affect life aspirations, life meaning, life satisfaction, happiness, and lifestyles. Several studies focus on the relations of these changes to the health status of individuals, or that of the population. Dissatisfaction with living circumstances, for example, contributes to alcohol abuse and other health risk behaviour, which are linked to ill-health, mortality and suicide rates. Indirect aspects of health, such as collapsing health care or hampered accessibility, are found in the majority of the post-socialist countries. In many of these countries, 'adequate' social status and corruption are becoming the prerequisites for access to medical assistance. The diverse effects of military or political crisis on various aspects of health are reported, often with special attention to adolescents and children. And finally, many quality of life or health-related studies observe the consequences of the nuclear power plant accident in Chernobyl, as well as the ongoing environmental, economical, and social tragedy in the Aral Sea Region, causing both deterioration in physical and mental health status.

Sense of Place in the Post-socialist Countries

'People of different cultures differ in how they divide up their world, assign values to its parts, and measure them' (Tuan 1977, 34). Beside the geographical, cultural, and historical characteristics, the process of post-socialist transition also contributes to the specific senses of places in these societies. In the reviewed literature (Krevs 2008) changes of personal attachments to national, regional, and local territories or places are the most frequently studied within this context. Among these changes, re-establishing a 'sense of Europe' bears specific connotations from the perspective of post-socialist societies. Wars, together with political and economic crises, caused – and still cause, intensive migratory processes all over the 'post-socialist region'. Based on these processes, the attachments to birthplace or home and nostalgic attachments to places, gain strength, and receive research attention. In the circumstances of a poorly functioning state, combined with economic decay and sometimes ethnic tensions, many local communities face a growing communal powerlessness, especially in rural areas. Such a 'hopeless' image of place predictably triggers emigration, especially of the young population.

This brief overview of the research mosaic specific to themes related to quality of life, health, and sense of place in post-socialist societies may serve as a frame for further discussion, as well as a pool of ideas for further research. Many of these topics that are currently studied alone by themselves may, in future, possibly be studied in a more relational, integrated fashion, allowing the relationships between quality of life, health and sense of place to be fully exposed.

Relating Quality of Life and Health to Sense of Place in the Post-socialist Countries

Many studies of sense of place implicitly also deal with some aspects of quality of life, such as happiness, or satisfaction with certain living conditions. Similarly, quality of life or health studies usually implicitly reflect the circumstances and perceptions of certain places. My intention is to present examples that show different relationships between quality of life, health, and sense of place as explicitly as possible. Despite the loose definitions used in the literature search, only a small number of studies have met these criteria. Some of them are briefly overviewed, with two presented in some detail.

Quality of Life and Sense of Place

Relations between suburban identity and quality of life are studied in a municipality in Zagreb's green belt (Lukić et al. 1997). The image of a green belt is related to a strong suburbanization process in 1990s. The study confirms the continuing high importance of unspoilt environment, life in a 'small community', and proximity of the city for perceived quality of life in, and for the attachment to, that municipality.

A contrasting example of urban neighbourhood attachment has been investigated in Petržalka, one of the biggest and the most densely populated neighbourhoods in Bratislava and Central Europe (Beňušková 2003). Its inferior reputation as general public opinion, and in the beliefs of older residents, is contrasted with more positive attachments that a new generation has to the neighbourhood, who identify with it as their home.

Several residential areas of Sarajevo have been evaluated from the perspective of selected population groups (Krevs and Drešković 2007). In the first phase, selected population groups – like pensioners, families with small children, and secondary school pupils, expressed their preferences about their 'good everyday living' conditions, such as proximity of a: kindergarten, school, park, coffeehouse, local vegetable market, and of health services. Based on the analysis of these living conditions, maps 'of suitability for everyday living' for each population group and for each studied residential area have been synthesized.

A Lithuanian study reveals an unsuccessful rehabilitation of the 'senses' of rural and urban places (Mincyte 2004). Since 1959, the reform of rural settlements in the Soviet Union strove to eliminate the distinctions in quality of life between urban and rural areas, and between the working-class and the peasants. It resulted in radical transformation of villages into 'agro-cities'. This action 'disembodied' or 'displaced' memories and material relics of the pre-World War II landscapes. Mincyte (2004) argues that reinstating historical land ownership rights will not lead to necessary improvements in urban and rural identities, or in the rural and urban quality of life.

Health and Sense of Place

The available studies that more explicitly link sense of place and health are divided into three groups, related to: (1) catastrophic events (2) displacement from home, and (3) to a very broad category of places, assigned either a healthy or unhealthy identity.

Catastrophes and Health

Catastrophic events, caused by natural or human agents, usually result in changes in perceptions of places, and in the health of the population. The accident in the nuclear power plant in Chernobyl in 1986 has demonstrated how a single event may harm the health of not only a local, but a regional or even a global population, and load the place (as well as the region and the country) with undesired characteristics and identities. Besides many studies related to the direct consequences of the radiation, some point out that also mental health and wellbeing have been damaged (Viinamäki 1995). A special aspect of the health effects of this disaster has been examined in a study (Havenaar et al. 1997) where two population samples have been compared six years after the accident – one from a seriously contaminated region in Belarus and another from a socio-economically comparable, but unaffected region in the Russian Federation. The results show how topophobic stereotypes, or negative senses of place, can result in long-term low levels of psychological well-being, health-related quality of life, and in illness behaviour in the exposed population. Significantly lower scores of self-reported health and higher use of medical services have been found out in the affected region, although at the time of the research no significant differences in global clinical indices of health were observed.

Displacement and Health

Leaving home for a longer period of time, voluntarily or under stressful circumstances, usually means at least a temporary tearing of many threads to the place one calls home. War in former Yugoslavia has caused a huge number of diverse types of displacements. A study of impacts of internal displacements of several hundred thousand people in Bosnia and Herzegovina on health and wellbeing (Kett 2005) points out the tragic situations these people find themselves in when considering returning to their homes. Beside the problems with belonging, housing, occupation, welfare, security and loss of communities, uncertainties and insecurities are further felt as a result of political pressures, experiences of violence, remembering the past, and issues of reconciliation.

Displaced refugee children during this war have been especially affected. In a Croatian study (Ajdukovic, M. and Ajdukovic, D. 1998), negative impacts on the psychological well-being of refugee children have been pointed out. Particular attention should have been devoted to children with traumatic experiences immediately prior to displacement. The children with poorer coping capacities and lacking a supportive family environment displayed high levels of stress-related symptoms throughout the entire refugee period, being at special risk for the development of further psychological difficulties.

Yet another type of displacement has been studied in the population of Lithuanian and Latvian seamen (Salyga and Juozulynas 2006). About a half of the population of the studied seamen have experienced psycho-emotional stress on the average 2.7–2.8 months from the beginning of the voyage. The occurrence of psycho-emotional stress had been mostly influenced by the work environment, resulting in increased visual strain and vibration. Depression occurred more frequently at sea than on shore, and has been associated with a disturbed working and resting regime resulting from time zone changes, and a disturbed regular sexual life.

Healthy and Unhealthy Places

Probably the most prominent examples of relation between sense of place and physical and mental health are represented by places seen as having therapeutic, spiritual, or faith related powers. Several 'therapeutic places' are used for regular medical treatments, like climatic health resorts or spas. Quite a few are used for 'alternative medicine' practices, or as places of religious pilgrimage. Among the latter the Catholic apparitional site in Medjugorje in Bosnia and Herzegovina, which holds a special position, receiving masses of pilgrims from all over the world. It also received scientific attention, illustrated in the medical monitoring of its visionaries (Joyeux and Laurentin 1986), or in the study of the effects of the sense of this place on the pilgrimages of former emigrants (Skrbis 2007). The latter study revealed sentimental feelings dedicated to Medjugorje by Croatian migrants in Australia. Going home, especially because it happens so seldom, is a heavily emotionally charged event in the life of a long-term emigrant. It affects the economic situation of the migrant's quality of life due to travel expenses and the costs of presents for the hosts at home. To some of the emigrants visiting home, a religious dimension is experienced to the secular journey. Some emphasize nationalistic views of Medjugoje as Croatian territory, through its connections with Croatian visionaries.[6] To some older migrants, to those with no home to go back to, and to those increasingly aware that they are tourists in their own homeland, Medjugorje becomes an 'imagined surrogate home' (ibid. 313).

A very direct influence on the characteristics of a place and the selected aspects of health are researched in studies of the impact of ambient temperature on mortality among the urban population in Skopje in the former Yugoslav Republic of Macedonia (Kendrovski 2006), and the association of rural setting and poorer quality of life in Parkinson's disease patients in Croatia (Klepac et al. 2007). Continental climatic seasonality in Skopje is resulting in seasonal mortality. Very cold winters cause a 7 per cent, and very hot summers a 13 per cent increase of daily mortality in Skopje. In the Croatian example, assessment of quality of life is used to evaluate 'total burden' of the illness instead of only the motor disabilities. Patients from rural areas had a significantly worse quality of life summary index score, and worse subscale scores, than their urban counterparts. The findings suggest the importance of considering the places where patients live as well as their socio-economic background, when attempting to achieve the best management of Parkinson's disease patient's needs.

Sense of Neighbourhoods and Quality of Life in Ljubljana

In a series of studies of perceptual spatial differentiation within the Municipality of Ljubljana (Kodre et al. 2000; Atelšek et al. 2001; Krevs, M. 2004; Kramar et al. 2007), changes in senses of neighbourhoods in Ljubljana, and their relation to quality of life have been researched. Among the research aims are: a better understanding of the spatial distribution and differences in the intensity of positive and negative senses of the neighbourhoods in Ljubljana; internal and external origin of the senses of neighbourhoods; temporal stability of these senses and of their relation to the quality of life, and eventual local mixing of different senses of individual neighbourhoods.

Collecting the Data

Face-to-face interviews, standardized during the whole series of studies, have been used to obtain the data. Respondents have been selected from each of the twenty-

---

6 'Virgin Mary has … revealed herself to Croatians … and she chooses to communicate her messages on what the participants in the research perceived as Croatian ethnic territoty' (Skrbiš 2007, 318).

seven neighbourhoods using quota sampling[7] (n=20 in 2001, 2004 and n=50 in 2007; total number of interviews n=540 and n=1350 respectively).

If the respondents need to express their perceptions about the neighbourhoods, these have to be recognizable, with some identity with and meaning to either the locals or 'outsiders'. In all these studies, neighbourhoods are constructed by researchers on the basis of former local communities, so that their identity can be expressed with 'generally recognizable' names, and are characterized by relative internal social-economical homogeneity. The recently introduced city districts are not suitable for such a study. They are quite bigger, named ambiguously, and many composed of very diverse local communities.

In the first group of questions respondents are asked to list up to three neighbourhoods that are: the most attractive for living, the least attractive for living, and the most unsafe.[8] Using more illustrative naming, the answers to the first question show 'love of a neighbourhood', answers to the second question 'hate of a neighbourhood', and answers to the third question 'fear of a neighbourhood'. The selection of up to three neighbourhoods instead of only one is supposed to make the respondent's decisions easier.

In the second group of questions the respondents are asked to give the main reasons for their answers to each of the questions from the first group.

Intensity of Senses of Neighbourhoods

The proportion of all the respondents that 'vote' for individual neighbourhood in relation to one of the three aspects of sense of place is a measure of the 'intensity' of sense of a neighbourhood. Positive attitudes to neighbourhoods are considerably more evenly spatially distributed, and consequently less spatially focused, than the negative ones. In the last study (Kramar et al. 2007) the neighbourhood most frequently selected as attractive for living was 'chosen' by 28 per cent of respondents, the most non-attractive neighbourhood for living by 49 per cent of respondents, and the most unsafe neighbourhood by 77 per cent of respondents. The intensity of these extreme senses of neighbourhoods, as well as the choice of the neighbourhoods, remained very stable since the first study in 2001. On the other hand 'less extremely perceived' neighbourhoods may change their 'rating' depending on recent events. For example building new high quality residential neighbourhoods may increase positive perceptions or alternatively, a criminal event receiving media attention may increase negative perceptions of a neighbourhood. The intensive negative perceptions have been found more stable than the intensive positive ones.

Personal Experiences, Clichés, and Local Mixing of Perceptions

As one of the studies suggests (Krevs 2004, 374), the positive perceptions are supposedly influenced by a wider range of factors, which are probably more often

7 The criteria for the quota sampling have been based on the age groups of the respondents and type of housing (e.g. multifamily high rise housing, single family houses) where they live.

8 A list and a map of all neighbourhoods have been handed to every respondent.

based on respondent's own experience. The negative perceptions may be based on a single criterion, possibly 'borrowed' from general public opinion and clichés. Selecting 'the worst' neighbourhoods is practically always 'pointing at others', while all the neighbourhoods, even 'the worst' by general opinion, are selected as 'attractive for living' at least by some locals.

Some of the neighbourhoods are hardly noticed by the respondents. They may lack positive or negative identity and could be referred to as 'placeless' (Relph 2002, 913). On the other hand, this is a matter of level of intensity of perception. Such neighbourhoods may be ranked just anywhere between 'placelessness' and the top three selections from a given aspect of sense of place.

Simultaneous study of all three aspects of senses of neighbourhoods shows that a diversity of perceptions is quite normal; intensive intrinsically opposite perceptions can be found for several neighbourhoods (like combination of 'love', hate' and 'fear' for neighbourhood Center, see Figure 10.2).

Relations to Quality of Life

Selected aspects of the quality of life are implicitly related to senses of the neighbourhoods in the first group of questions. The respondents need to tell where

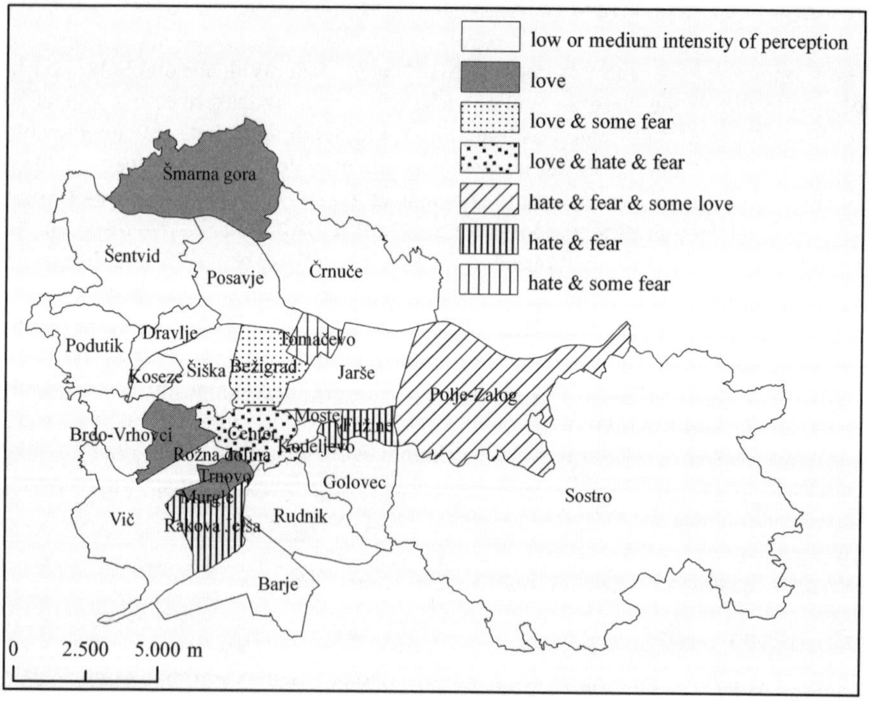

Figure 10.2 Topophilia and topophobia of the neighbourhoods within the Municipality of Ljubljana

Source: Based on Krevs (2004, 377).

they think 'the good, the bad, and the dangerous' life activities are located within the Municipality of Ljubljana. Their spatial distribution of quality of life does not necessarily coincide with the assessments of quality of life that we could get from the local population, or correlate with "objective quality of life" indicators.

In the answers to the second group of questions the respondents explain their conceptualizations of the three aspects of quality of life. The most frequent factors influencing positive perceptions of neighbourhoods and the quality of life found there, are: proximity to green areas, clean, attractive and peaceful residential environment, accessibility of services, traffic, and public and personal safety. The negative perceptions of neighbourhoods are most frequently influenced by: an environment which is noisy, polluted, unattractive, or densely populated, high rise housing, accessibility to services or to centre of the city, national structure of population, and the lack of public and personal security. The perceptions of unsafe neighbourhoods are most frequently based on: criminal activity, violence, illegal drug trafficking, rumours, news in the media and stereotypes, a large immigrant population, national structure of the population, and negative personal experiences.

Moderate or low correlations are found between the 'intensity' of senses of neighbourhoods and selected indicators of circumstances of living, such as income, national heterogeneity, and size of residences. The positive perceptions of the neighbourhoods tend to be more related to materialistic values and correlate with spaciousness of the residence and the economic status of the residents. The negative perceptions seem to be more related to prevailing stereotypes, including nationalistic views, and correlate with higher share of 'Non-Slovenian' population and smaller size of residences.

If the intensities of senses of neighbourhoods are combined with the 'types of circumstances of living' from another study (Krevs 2002), the most positively perceived neighbourhoods are either those with a population of higher socio-economic status in the central part of the city, or a suburban area at the northern edge of the municipality, below a hill named Šmarna gora – a popular destination of family excursions at weekends. The majority of the most negatively perceived neighbourhoods are not far from the city centre. These are socially deprived areas characterized by mostly unfavourable living conditions such as: illegally constructed single-family houses, inadequate infrastructure, and poor accessibility of services. However, the most 'hated' and 'feared' neighbourhood, named Fužine, is a high rise housing neighbourhood constructed in the 1980s, with modern infrastructure, access to supplies, medical and other services. The negative perceptions are obviously based on the high density of population, and especially on the high share of immigrants from other republics of former Yugoslavia. The 'fear' of the neighbourhood is therefore constructed based on the perceived association between the ethnic structure of the population and criminality, or more general feelings of safety which are only partly supported by the results of a detailed study of criminality in Ljubljana (Lampič 2004).

Sense of Mediterranean and Health in Slovenia

Modern fascination with the Mediterranean is mainly based on the belief that the Mediterranean lifestyle is healthier than the lifestyles in continental Europe, leading to lower cardiovascular disease mortality.

A series of studies (Staut et al. 2005; Staut and Kovačič 2006; Staut 2007) examines the relation between Mediterraneity and health in Slovenia. In the first step 'sense of Mediterranean' has been researched using the method of mental mapping. Mental maps have been drawn by 193 respondents. Digitalization of mental maps allowed analysing their accumulated spatial overlies (Figure 10.3a), resulting in a fuzzy representation of the aggregated spatial sense of the Mediterranean (Figure 10.3b).

Specificities of the Mediterranean climate, vicinity of the sea, type of vegetation, character of residents – or more broadly, the way of life are the aspects reported to be taken into account while drawing the mental maps. Relation to health, although quite rough and indirect, is shown by age-standardized cardiovascular disease mortality rate, which supports the stereotype about 'more healthy' Mediterranean areas in comparison to the rest of the country (Figures 10.3c and 10.3d). Another indirect link between the Mediterranean and cardiovascular diseases is supposed to be the rapidly growing olive oil and red wine production. The diminishing difference in the cardiovascular disease mortality rate between the Mediterranean and other areas in Slovenia (Figure 10.3d) might be at least partly related to the spread of the Mediterranean lifestyle to other parts of the country. But this is just an assumption to be checked in future studies.

Discussing Sense of Place, Quality of Life and Health in Post-socialist Countries

Researching both the objective or subjective aspects of quality of life has quite a tradition in post-socialist societies. Further, studies of health in these countries are quite widespread, often carried out in association with the research on quality of life. Comparatively, the quality of life studies incorporate more geographical aspects than do the health studies.

Senses of places have been investigated very sporadically in post-socialist societies. Traditional regional geographical studies do emphasize the geographical uniqueness of these societies, aiming to present their intrinsic character and, by so doing, actually transforming areas into places. But these can hardly be seen as studies of senses of places, since the ascribed characteristics usually tell very little about how, and how intensively if at all, people perceive and experience places.

Research of the inter-relation between quality of life, health, and sense of place is therefore only beginning in post-socialist countries. The selected examples are mostly focused on the western part of the region that comprise these countries. This geographical bias is reflected in some of the themes, such as the influence of Mediterranean way of life on health, effects of the displacement of seamen on their psychological-emotional stress and health, or the link between the perceptions of neighbourhoods and quality of life, which might easily be found on the research

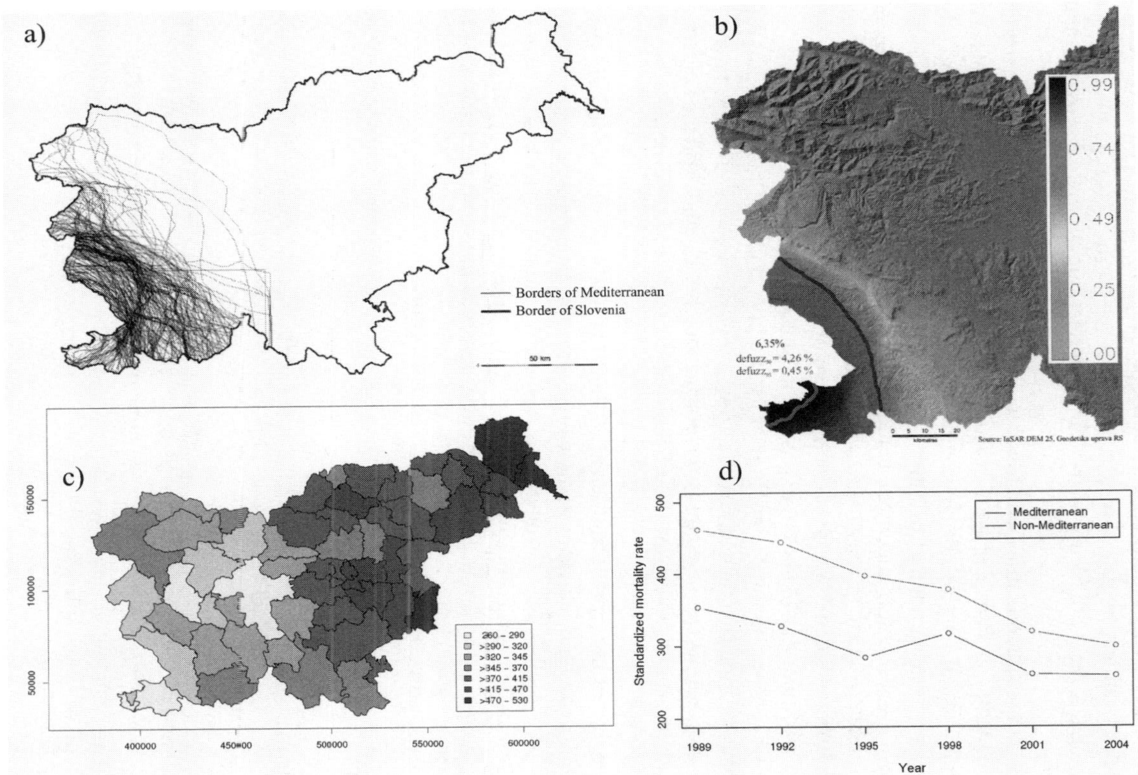

Figure 10.3 Mental mapping of perceived Mediterranean

Sources: Images and mental maps: Staut and Kovačič (2006); Staut (2007); data on mortality: Institute of Public Health of the Republic of Slovenia.

agenda of any Western European country. On the other hand, some of the overviewed themes and findings seem very specific, like the relation between the religious visitation to an apparition site in Medjugorje and nationalistic feelings of Croatians, effects of displacements on health due to the war in former Yugoslavia, or the proximity of *buregdžinica* or *ćevabdžinica* among factors of 'good everyday living" of secondary school pupils in Sarajevo.[9] Still, these examples make a very narrow and biased selection. Several studies pointed out in the overview (Krevs 2008) that focus either on quality of life, health, or senses of places in post-socialist societies, suggest there are plenty of specific inter-relations yet to be discovered and researched.

In the post-socialist countries with lower levels of objective quality of life, research is more frequently preoccupied with existential problems such as solving problems of economic, social, political, or environmental survival. Results of the analysis in this chapter support the findings of other studies (e.g. Fahey and Smyth 2004; Hayo 2007), specific to economic conditions being more directly related to other aspects of objective quality of life, happiness, and life satisfaction in post-socialist societies than in Western Europe or in the World. Economic development may therefore be expected to increase quality of life and health in post-socialist countries.

And how can we support the arguments for the importance of the linking of the sense of place to quality of life or health in post-socialist societies? There is no easy way at the moment. Even in the highly developed societies of the World the evidence is accumulating quite slowly. Much of the evidences may not be transferable to post-socialist societies. Besides, this evidence does not address many of the specific issues of quality of life, health, and sense of place in these societies. Additional case studies are necessary. Such case examples will most probably not be gathered in a systematic way to allow expressing generalized expectations such as those related to the influences of economic development on quality of life or health; they will likely be mainly focused on answering concrete questions, related to certain places, within certain contexts.

References

Ajdukovic, M. and Ajdukovic, D. (1998), 'Impact of displacement on the psychological well-being of refugee children', *International Review of Psychiatry* 10(3): 186–195.

Alber et al. (2004), *Quality of life in Europe* (Dublin: European Foundation for the Improvement of Living and Working Conditions).

Atelšek, G. et al. (2001), *Topofobne in topofilne predstave o soseskah v Mestni občini Ljubljana*, Compendium of students' research reports and database (Ljubljana: Department of Geography, Faculty of Arts, University of Ljubljana).

9 Two examples of traditional 'fast food' restaurants in Bosnia and Herzegovina, and elsewhere in the Balkan region; in *burekđinica* a phyllo pastry, usually filled with cheese or meat is served, and in *ćevabdžinica* a dish of grilled minced meat is served on a plate or in a flatbread.

Beňušková, Z. (2003), 'Lebensqualität in der Wohnsiedlung Petržalka in Bratislava', *Österreichische Zeitschrift für Volkskunde* 106(2) 157–164.

Constantini, V. and Monni, S. (2005), 'Sustainable Human Development for European Countries', *Journal of Human Development* 6(3): 329–351.

EIU – The Economist Intelligence Unit (2005), *The Economist Intelligence Unit's Quality of Life Index, The World in 2005* <http://www.economist.com/media/pdf/QUALITY_OF_LIFE.pdf>, accessed 8 June 2007.

Eyles, J. (1985), *Senses of Place* (Warrington: Silverbrook Press).

Fahey, T. and Smyth, E. (2004), 'Do subjective indicators measure welfare? Evidence from 33 European societies', *European Societies* 6(1): 5–27.

Fajth, G. (2000), 'Regional monitoring of child and family well-being: UNICEF's MONEE project in CEE and the CIS in a comparative', *Statistical Journal of the UN Economic Commission for Europe* 17(1): 75–100.

Franz Sladovic, B. (2004), 'Predictors of behavioural and emotional problems of children placed in children's homes in Croatia', *Child and Family Social Work* 9(3): 265–271.

Hayo, B. (2007), 'Happiness in transition: An empirical study of Eastern Europe', *Economic Systems* 31(2): 204–221.

Havenaar, J. et al. (1997), 'Health Effects of the Chernobyl Disaster: Illness or Illness Behavior? A Comparative General Health Survey in Two Former Soviet Regions', *Environmental Health Perspectives Supplements* 105(6): 1533–1537.

International Living (2007), *Quality of Life Index 2007* <http://www.internationalliving.com/issues/2007/2007_article.html>, accessed 5 June 2007.

Joyeux, H. and Laurentin, R. (1986), *Etudes medicales et scientifiques sur les Apparitions de Medjugorje* (Paris: OEIL).

Kendrovski, V.T. (2006), 'The impact of ambient temperature on mortality among the urban population in Skopje, Macedonia during the period 1996–2000', *BMC Public Health* 6: 44.

Kett, M.E. (2005), 'Internally Displaced Peoples in Bosnia-Herzegovina: Impacts of Long-term Displacement on Health and Well being', *Medicine, Conflict and Survival* 21(3): 199–215.

Klepac, N. et al. (2007), 'Association of rural life setting and poorer quality of life in Parkinson's disease patients: a cross-sectional study in Croatia', *European Journal of Neurology* 14(2): 194–198.

Kodre, L. et al. (2000), *Življenjske razmere v Ljubljani*, Compendium of students' research reports (Ljubljana: Department of Geography, Faculty of Arts, University of Ljubljana).

Kramar, N. et al. (2007), *Privlačnost sosesk v Mestni občini Ljubljana*, Compendium of students' research reports and database (Ljubljana: Department of Geography, Faculty of Arts, University of Ljubljana).

Krevs, M. (2002), 'Geografski vidiki življenjske ravni prebivalstva Ljubljane', *Geografija Ljubljane* (Ljubljana: Department of Geography, Faculty of Arts, University of Ljubljana).

—— (2004), 'Perceptual spatial differentiation of Ljubljana', *Dela* 21: 371–379.

—— (2007), 'Quality of life and health in post-socialist countries". *Dela* 30: (forthcoming).

Krevs, M. and Drešković, N. (2007), *Višekriterijsko vrednovanje ugodnosti lokacija u gradu Sarajevo sa aspekta izabrane grupe stanovnika*, Research report (Lubljana: Oddelek za geografijo, Filozofska fakulteta, Univerza v Lubljani).

Lampič, B. (2004), 'Kriminaliteta kot vse pomembnejši dejavnik kakovosti bivanja v ljubljanski urbani regiji' [Criminality as an increasingly important factor of the quality of living in the Ljubljana urban region'], *Dela* 22: 129–140.

Lukić, A. et al. (1997), 'Suburbanizacija i kvaliteta življenja u zagrebačkom zelenom prstenu: primjer općine Bistra', *Hrvatski geografski glasnik* 67(2): 85–106.

McDowell, L. (ed.) (1997), *Undoing Place? A Geographical Reader* (London: Arnold).

Mincyte, D. (2004), 'Displacement of Memory through Reforms of Rural Settlements in Lithuania', *Memory and the Present in Postsocialist Cultures*, Soyuz: The Research Network for Postsocialist Cultural Studies" <http://www.uvm.edu/~soyuz/frameset.html>, accessed 5 June 2007.

NEF – The new economics foundation (2006), *The Happy Planet Index* <http://www.happyplanetindex.org/>, accessed 8 June 2007.

Post-Socialist Geographies Research Group (home page) within Royal Geographical Society (with Institute of British Geographers) <http://www.psgrg.org.uk/>, accessed 5 June 2007.

Relph, E. (2002), 'Place', in Douglas et al. (eds) *Companion Encyclopedia of Geography. The Environment and Humankind* (London: Routledge).

Salyga, J. and Juozulynas A. (2006), 'Association between environment and psycho-emotional stress experienced at sea by Lithuanian and Latvian seamen', *Medicina* 42(9): 759–769.

Skrbis, Z. (2007), 'From Migrants to Pilgrim Tourists: Diasporic Imagining and Visits to Medjugorje', *Journal of Ethnic and Migration Studies* 33(2): 313–329.

Soyuz: The Research Network for Postsocialist Cultural Studies (home page), Post-Communist Cultural Studies Interest Group of the American Anthropological Association (AAA), also recognized as an official unit of the American Association for the Advancement of Slavic Studies (AAASS) <http://www.uvm.edu/~soyuz/frameset.html>, accessed 5 June 2007.

Staut, M. (2007), 'Contextualizing the idea of the "healthy Mediterranean" in Slovenia: diverging processes in health related cultural Practices', in 2007 annual meeting abstracts: April 17–21, 2007 (Washington, DC: The Association of American Geographers), 606.

Staut, M. and Kovačič, G. (2006), *Slovensko Sredozemlje in življenjska raven povezana z zdravjem prebivalstva,* Research report (Science and Research Centre, University of Primorska, Koper, Slovenia).

Staut, M. et al. (2005), 'Prostorsko dojemanje Sredozemlja v slovenski Istri: analiza s pomočjo teorije mehkih množic', *Annales. Series historia et sociologia* 15(2) 427–436.

Tuan, Y.F. (1974, Morningside Edition 1990): *Topophilia, A Study of Environmental Perceptions, Attitudes, and Values* (New York: Columbia University Press).

—— (1977, 9th reprint 2002), *Space and Place, The Perspective of Experience* (Minneapolis: University of Minnesota Press).

UNDP – United Nations Development Programme (2006), *Human Development Report 2006* <http://hdr.undp.org/hdr2006/statistics/>, accessed 8 June 2007.

United Nations Economic Commission for Europe – UNECE (2004), 'Chapter 7 – Poverty in Eastern Europe and CIS', *Economic Survey of Europe*, 2004 Issue 1, 163–76, <http://www.unece.org/ead/pub/041/041c7.pdf>, accessed 5 June 2007.

Viinamäki, H. (1995), 'The Chernobyl accident and mental wellbeing: A population study', *Acta Psychiatrica Scandinavica* 91(6): 396–401.

Chapter 11

Environment and Health: Place, Sense of Place and Weight Gain in Urban Areas

Paula Santana and Helena Nogueira

1. Introduction

Questions about the interrelationships between the socioeconomic environment of a neighbourhood, its physical characteristics and the health status of its inhabitants have been attracting interest in recent years from researchers in both the academic and political sphere. Today, there is a strong body of evidence linking population health and quality of life to the social and physical settings in which people caring out their daily activities (Young et al. 2004). Mortality, respiratory and coronary heart diseases, depression and other mental conditions, obesity, and general health status, as well as health-related behaviours (tobacco consumption, diet, physical activity) has been consistently related with the local environment (Diez-Roux et al. 2001; Lochner et al. 2003; Van Lenthe et al. 2005; Veenstra et al. 2005; Nogueira 2006; Day 2007).

However, the challenge remains of how best to conceptualize and operationalize those aspects of the local environment that have an impact upon health. In the UK, Macintyre and colleagues have proposed a framework for understanding neighbourhood as a "local opportunity structure" (Macintyre and Ellaway 1999; Macintyre et al. 2002) that people can use, or fail to use. Some of the aspects belonging to that structure that should be given a major place in health research include air and water quality, the availability of healthy environments at home, work and play (e.g. adequate housing, safe play areas for children), the proximity of facilities and services to provide support for people in their daily lives (e.g. food shops, transport, policing, street cleaning) and the sociocultural features of localities and communities (e.g., heritage sites, community integration, collective efficacy, social capital).

According to Andrews and Kearns (2005), place actively shapes health – individual health and health experience are both constrained and enabled by the character of place. Furthermore, health can also shape the image and identity of "everyday" places; thus, place is shaped by the health of residents. This theoretical perspective, originating from cultural geography, strengthens both the role of place and sense of place in health and the role of health in places.

However, a sense of place is even more elusive than place itself. A sense of place appears to result from a complex process, involving objective and subjective dimensions, including the physical and socioeconomic characteristics of the environment, emotional attachment to the neighbourhood and community, levels of interaction between members of the community, and formal participation or involvement in neighbourhood and community organizations (Young et al. 2004). These factors are interrelated, because the physical features of the neighbourhood may influence the development of social ties.

The general deterioration of place and sense of place may be triggered by several processes together. Poor disadvantaged neighbourhoods tend to be avoided, and those that live in them may not intend to stay there for long. Consequently, houses, public spaces and other physical features of the neighbourhood are not properly maintained, and decay sets in. Previous facilities and amenities may disappear and crime may become more prevalent (Van Lenthe et al. 2005). Neighbourhoods that have lost their attractiveness, accessibility and proximity to facilities and that are unsafe are not conducive to the establishment of social contacts and to the development of social interactions. Social capital, social cohesion and social support declines (Young et al. 2004). The place may become unpleasant and this will have a detrimental effect on health.

1.1. Neighbourhood, Physical Activity and Obesity

Obesity has reached epidemic proportions in almost all developed countries (Backett-Milburn et al. 2006). Weight gain research has mainly focused on individual factors, such as biological, psychological and behavioural aspects (consumption of fast foods, television viewing, low levels of leisure and physical inactivity) (Poortinga 2006). Given that no significant mutations or changes in human anatomy have occurred in the last few decades, it is probable that the primary etiology for the weight gain is neither genetic nor physiological (Cohen et al. 2006). Thus, in spite of the genetic influence on obesity, the most recent research (Kim et al. 2006; Poortinga 2006) points to environmental factors, and their potential obesogenicity, that is the extent to which they could promote caloric intake and/or discourage the expenditure of energy in routine physical activity. It is important, therefore, to identify exactly what those factors are.

Many researchers (Frank and Engelke 2001; Calthorpe and Fulton 2001; Wilkinson and Marmot 2002; Jackson 2002; Jochelson 2004; Giskes et al. 2005; Kim et al. 2006; Poortinga 2006) have found direct and indirect links between the neighbourhood environment and BMI. According to Poortinga (2006), this research focuses on the contextual effects of:

1. Access to amenities: food shops, health services, sports facilities, for example
2. Physical features of the environment: presence of green spaces, cycle paths, and the degree of urbanization
3. Neighbourhood reputation: reflected in feelings of safety, crime figures, and so on

4. Aesthetics: the role of general attractiveness of the neighbourhood
5. Social organization and collective functioning: the protective effect of social capital and social support.

All of these factors appear to be associated with BMI, a standard measure of obesity. Factors related to urban sprawl (for example, areas with low population density and low street connectivity), land use (living in homogeneous land use areas), location of facilities (sports and leisure centres, public services, shops, parks and green spaces located far from area of residence) and community perceptions (unsafe and unpleasant environments) could create "unwalkable" communities and are related to a prevalence of excess weight and obesity. In addition, there is some support for the hypothesis that social capital and social support are beneficial for a wide range of physical activity outcomes, like walking, cycling and sports engagement (Lindstrom et al. 2003; Poortinga 2006). Cohen et al. (2006) suggest that collective efficacy (concept closely associated with social capital) is related to BMI, although the mechanisms that link neighbourhood collective efficacy to weight status were unclear. Collective efficacy enables collective action and strength social interactions, and this may affect the sense of belonging to a community.

"Walkability" could be understood as the extent to which the characteristics of the built environment and land use may or may not be conducive to residents walking for leisure, exercise or recreation, to access services, or to travel to work (Leslie et al. 2006). In USA, inhabitants of dense counties weighed less and had a lower BMI than their counterparts residing in sprawling counties (Pendola and Gen, forthcoming). Mixed land use leads to "high-walkability" communities, and people residing in those communities had, on average, a lower BMI than people residing in "low-walkability" communities. As well as the physical, the social environment can also provide, or fail to provide, opportunities to engage in particular behaviours, which may reduce or produce stress, as well as place constraints on individual choices.

Kim et al. (2006) stressed the role of area deprivation, urban sprawl, social capital and safety and security on levels of physical activity. Poortinga (2006) highlights the links between social capital, social support, physical activity, obesity and the overall health of residents. Social capital and support stimulates various positive health behaviours, including physical activity, and exerts some measure of social control over deviant health behaviours, such as smoking and alcohol abuse (Kawachi and Berkman 2000). Individuals living in disadvantaged neighbourhoods, both social and physical, are at increased risk of engaging in unhealthy behaviours, such as physical inactivity, often as a response to a stressful and hazardous environment (McNeill et al. 2006). In Australia, Young et al. (2004) report a strong interrelationship between the sense of belonging to a neighbourhood, feeling safe and physical activity.

2. Objectives

The purpose of this paper is to examine the contribution of the local environment (physical and social), and personal attributes to the risk of weight gain and obesity in neighbourhoods of the Lisbon Metropolitan Area (LMA). To achieve this, we planned to: (1) develop indicators to measure dimensions of the local physical and immaterial environment; (2) link individuals to their neighbourhoods, through the characterization of each individual personally and according to features of their place of residence; (3) establish associations between the local environment and residents' body mass index (BMI).

The next section describes the data and methods used, followed by the main results achieved. Finally, in the last part of the chapter, possible explanations are suggested, some conclusions are drawn and key proposals are made for changes that could lead to improved health results.

3. Data and Methods

The LMA comprises nineteen municipalities, 216 wards and over two and a half million inhabitants. The sampling universe consisted of the residents of the LMA. A representative sample of 7,669 individuals, corresponding to 143 neighbourhoods, was collected by trained interviewers (Portuguese Health Survey – NHS – 1998/99) (Figure 11.1).

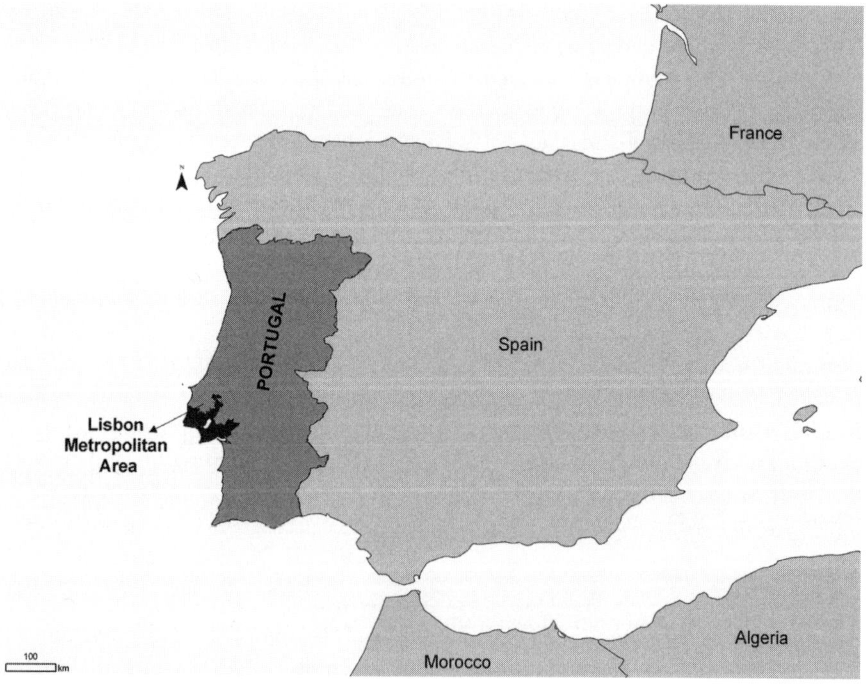

Figure 11.1 Map of the LMA

3.1. Outcome Measure

Weight gain was assessed through body mass index (BMI), calculated from the respondent's reported height and weight. The BMI variable was categorized into 3 groups: those who have a normal weight, defined as having a BMI ≥ 18.5 and <25; those who are overweight, defined as having a BMI ≥ 25 and <30; those considered obese, having a BMI ≥ 30.

3.2. Predictor Variables

At the *individual level*, some key characteristics related to weight gain were collected from the NHS. These were: demographic (age, gender), economic activity (employed, unemployed, students, retired and housewife) and behavioural (physical activity, diet). All individual variables were included as dummy variables, except age, which was included both as continuous and as age-squared in order to model the curvilinear relationship between age and BMI. In fact, it was expected that BMI would increase with age through middle adulthood and then decline in old age (Robert and Reither 2004).

At the *neighbourhood level*, we created eight new indicators in order to measure local environmental dimensions possibly related to weight gain and obesity. Data was obtained from a large range of routine and non-routine sources. These include the National Institute of Statistics (INE), local authorities, voluntary and public sector agencies (ministry of health, local and municipal police, the Portuguese social security), commercial organizations and others (the national institute of car insurance and yellow pages). We tried to generate true and specific contextual measures, beyond aggregate measures. According to Cummins et al. (2005), the latter could be problematic, since they generally do not reflect the underlying social process that links social and physical environment to health outcomes. Addressing the challenge of accurately reflecting a range of specific features of the local environment that may influence a particular health outcome like BMI, some ecological measures were created using statistical methods:

1. An index of multiple deprivation: occupation (men's unemployment and people with non-qualified jobs); living conditions (percentage of people living in shanties);
2. Factors or components (7) underlying the social, physical and environmental dimensions.

The seven contextual indicators addressed in the present paper were considered to adequately support research into the role of the environment on BMI and obesity. These are (see Table 11.1):

1. Linking social capital
2. Bonding social capital
3. Crime and policing
4. Preventive health services available

5. Public transport accessibility/car use
6. Sports facilities
7. Residential environment (housing adequacy).

In relation to social capital we followed an approach developed by Szreter and Woolcock (2004). According to these authors, social capital is a multidimensional concept, which can thus be unpacked into bonding, linking and bridging social capital. Bonding social capital involves social support provided by horizontal ties between members of social networks, and can be compared to the concept of social cohesion. Social cohesion and positive social interactions can strength identities and sense of belonging to a community. Linking social capital refers to vertical interactions across power or authority structures in society, stressing the effective mobilisation of political institutions. Bridging social capital, not measured in this study, assumes inequality and highlights the need for justice, solidarity and respect across the social spectrum.

3.2.1. Creating New Contextual Variables

1. Multiple deprivation index: In order to highlight the link between multiple deprivation and BMI, we used a composite index of area deprivation. This was created by selecting three census variables related to socioeconomic characteristics: occupation (1. unemployment; 2. unskilled employment) and living conditions (3. percentage of individuals living in slum conditions). Following the method of Carstairs and Morris (1991), each variable was standardized to give a population-weighted mean of zero and variance of one (z-score method). The deprivation index results from the sum of the new standardized variables (Nogueira and Santana 2005).
2. Components underlying the dimensions of the local physical and immaterial environment: We collected a large amount of contextual data (84 variables), in non-routine data sources, which were then assigned to the previous seven environmental dimensions or constructs (see Table 11.1). To explore and reduce these data, Principal Component Analysis (PCA) was performed. The aim of PCA was to reduce the number of variables potentially used in statistical analysis. All components were rotated using varimax rotation to maximise factor loadings and rejected when considered irrelevant using Kaiser's criterion. Those variables with a low loading onto components were discarded from the construct. The number of variables entering into each PCA was systematically modified several times in order to generate a single, strong component in each domain (Nogueira 2006). Overall, seven components or factor scores were extracted, with thirty variables.

Table 11.1 describes the seven components generated with PCA. It also shows two reliability measures: the Cronbach-alpha score and the standardized Alpha. Alpha scores run between zero and onc, with higher values indicating greater reliability. In this case, standardized Alpha scores ranged from 0.51 to 0.92. These high values show that variables within each component are strongly related, which gives us confidence about the internal consistency of the extracted components and their ability to measure the latent contextual domains. Correlations between factor scores were generally low, suggesting single, strong and unambiguous components, with the ability to reliably capture something unique about the local environment.

Table 11.1 Summary of neighbourhood constructs

Cronbach/ Standardized Alpha	Constructs/ Dimensions	Items in each Component
0.61/0.92	Crime and Policing	No. of crimes against the person per 1000 pop. No. of crimes against property per 1000 pop. No. of crimes against society per 1000 pop. No. of crimes against state property per 1000 pop. No. of other crimes per 1000 pop. No. of road crimes per 1000 pop.
0.68/0.73	Sport	No. of indoor pools No. of sports grounds No. of gymnasiums Tennis courts, riding clubs and golf courses Fitness circuits, athletics tracks and skating rinks
0.55/0.55	Accessibility	Distance from the parish[a] to the Town Hall by public transport Distance from the parish[a] to the town centre by public transport Percentage of people using private transport
0.80/0.85	Linking social capital	Percentage of abstentions 2001 (participation in parish elections) Percentage of abstentions 2002 (participation in national elections) Percentage of abstentions 2005 (participation in national elections)
0.27/0.51	Bonding social capital	No. of clubs for recreational or sports activities No. of local newspapers Local newspaper circulation per inhabitant
0.75/0.86	Preventative health services available	No. of pharmacies No. of clinical testing laboratories No. of x-ray services No. of echography services No. of CT scan services
0.78/0.92	Housing adequacy	No of dwellings without electricity No of dwellings without water No of dwellings without toilet No of dwellings without sewerage systems No of dwellings without bath

[a] "Parish" (small administrative area, and census ward)

3.3. Analysing the Influence of Individual and Contextual Characteristics on BMI – a Logistic Regression Analysis

As discussed above, the BMI was divided into three categories. Modelling BMI categories involves ordered categories from which the BMI_j category is chosen and predictors are related to individual and contextual characteristics.

The ordered logit model depends upon the idea of cumulative probability. The cumulative probability of BMI_{ij} is the probability that the *i*th individual is in the *j*th or higher category:

$$\text{logit}\left(BMI_{ij}\right) = \log\left(BMI_{ij}/1 - BMI_{ij}\right)$$

The ordered logit model simply models the cumulative logit as a linear function of independent variables:

$$\text{logit}\left(BMI_{ij}\right) = \alpha_j - \beta x_i$$

There is a different intercept for each level of the cumulative logit; this means that each α_j indicates the logit of the odds of being equal to or less than category *j* for the baseline group (when all independent variables are zero). Thus, these intercepts will increase over *j*. These intercepts are sometimes referred to as cut-points. The β does not vary with the level of the cumulative logit. Also note that β is subtracted rather than added. The β shows how a one-unit increase in the independent variable increases the log-odds of being higher than category *j* (due to the negative sign) and it is assumed that a one-unit increase affects the log-odds the same way regardless of which cut-point is being considered. The odd ratios presented in the results were calculated from the log-odd (β).

We started with a simple model, considering individual factors as independent variables. In this model, we selected the variables with a conventional significance of $p \leq 0.05$. At a second stage, we developed this model by entering each of the ecological variables that were selected if they were significant ($p \leq 0.05$). Six ecological variables were selected and combined in a full model. However, when all contextual predictors are entered simultaneously, several lose their statistical significance. In an attempt to retain the influence of the largest possible number of variables, two models were achieved: Model 1 – which includes individual factors, area deprivation score, residential environment and preventive health services available; and Model 2 – which includes individual factors, linking social capital (political participation) and availability of sport facilities.

4. Results

The models displayed in Table 11.2 show the influence of individual and contextual characteristics upon the odds of reporting excess weight.

In the two models, the coefficients (not shown) of individual variables were very similar, showing a similar influence on BMI. Generally, women have a lower risk

Table 11.2 Logistic regression models of overweight BMI

	Model 1	Model 2
Individual variables		
Gender	0.78* (0.71; 0.86)	0,77* (0,7; 0,85)
Age	4.74* (4.03; 5.57)	4,73* (4,03; 5,56)
Age-squared	0.99* (0.99; 1.0)	0.99* (0.99; 1,0)
Practice of activity	0.66* (0.6; 0.73)	0.65* (0.59; 0.72)
Economic activity		
Employed	1.00	1.00
Unemployed	1.01 (0.8; 1.02)	1.01 (0.85; 1.21)
Retired	1.17** (1.01; 1.36)	1.19** (1.02; 1.38)
Student	0.79 (0.57; 1.09)	0.78 (0.56; 1.07)
Housewife	1.22** (1.03; 1.43)	1.24** (1.05; 1.46)
Contextual variables		
Deprivation score	1.03* (1.01; 1.05)	
Housing adequacy – Housing quality (indoor)	1.2* (1.1; 1.31)	
Preventive health services available	0.96** (0.92; 1.00)	
Sports facilities		0.95** (0.91; 0.99)
Linking social capital		1.07** (1.01; 1.13)
Model fitting information:	Log Likelihood Ratio =1007.6**	Log Likelihood Ratio =1023.1**
Pseudo R-square	Cox and Snell =0.123 Nagerkerke =0.143 McFadden =0.066	Cox and Snell =0.125 Nagerkerke =0.145 McFadden =0.067

Note: OR for age based on a 10-year change; *p≤0.01; **p≤0.05.

of excess weight (22 per cent less likely to be overweight; OR=0.78, 95 per cent CI=0.71–0.86). With regards to age, the positive age coefficient and negative age-squared coefficient demonstrate the positive association between age and BMI in youth and adulthood – the odds of individuals reporting excess weight or obesity are five times more likely for each additional ten years – followed by a slight decrease in old age (OR=4.74, 95 per cent CI=4.03–5.57). BMI rises with age through middle adulthood and then declines in old age.

Economic activity is also a significant factor, although not for all the categories specified. In relation to employment, the odds of having excess weight are higher for retired people (19 per cent; OR=1.17, 95 per cent CI=1.01–1.36) and housewives (24 per cent; OR=1.22, 95 per cent CI=1.03–1.43); being unemployed or a student is not a significant determinant of excess weight.

Regarding behavioural factors, the influence of diet (consumption of vegetables or fruit) did not show any statistical significance on BMI. On the other hand, physical activity is a significant factor for diminishing the odds of individuals reporting excess weight or obesity; exercise reduces the odds of excess weight by 34 per cent (OR=0.66, 95 per cent CI=0.6–0.73).

In relation to contextual variables, accessibility to public transport/vehicle dependence, civic involvement (bonding social capital) and crime did not reveal any statistical significance and were removed from the models. The other five contextual variables (area deprivation, basic housing conditions, availability of preventive health services and sports facilities, political participation) were combined into two models (see above).

According to the models achieved, the highest risk of excessive weight was observed for:

1. Model 1 – people who lived in neighbourhoods with higher levels of material deprivation (3 per cent more likely to be overweight; OR=1.03, 95 per cent CI=1.01–1.05), poor housing (21 per cent more likely to be overweight; OR=1.2, 95 per cent CI=1.1–1.31) and poor availability of preventive health services (4 per cent more likely to be overweight; OR=0.96, 95 per cent CI=0.92–1.0)
2. Model 2 – people living in neighbourhoods with lack of sport facilities (5 per cent more likely to be overweight; OR=0.95, 95 per cent CI=0.91–0.99) and lower levels of linking social capital (political participation, 7 per cent more likely to be overweight; OR=1.07, 95 per cent CI=1.01–1.13).

5. Discussion and Conclusions

Physical inactivity is a growing public health problem associated with increased risk of obesity and heart disease, diabetes, high blood pressure and some cancers (Lopez-Zetina et al. 2006). The World Health Organisation (Davies 2002) recommends about 30 minutes of moderate physical activity (PA), such as that provided by walking and cycling, at least five days per week. Walking and cycling, both as recreation and as a means of daily transport can prevent weight gain and reduce the risk of obesity-related diseases (Jackson 2002).

In this chapter, individual and ecological data have been modelled simultaneously to account for the odds of reporting overweight and obese BMI. A range of contextual constructs were created and tested in a logistic regression framework. The results provide evidence of the protective effects of some environmental features on the likelihood of weight gain. With adjustments for a number of individual variables (gender, age, socioeconomic and physical activity), higher levels of deprivation, lack of preventive health services, poor housing, lack of sports facilities and lower levels of linking social capital were associated with higher relative odds of excessive weight. Housing, linking social capital and sports facilities show the strongest influence on BMI (respectively 21 per cent, 7 per cent and 5 per cent).

The results achieved for LMA are in agreement with previous studies (Robert and Reither 2004; Van Lenthe et al. 2005; Poortinga 2006). Individuals living in more vulnerable areas are more likely to report excessive weight than their counterparts living in more affluent areas. Similarly, a growing body of evidence suggest that physical and mental health problems, such as obesity, are related to inadequate housing (Fullilove and Fullilove 2000).

Some of the new ecological variables allow us to demonstrate that certain features of the local environment may contribute to the growing rates of physical inactivity, excess weight and obesity. The topics researched here clearly provide a meaningful framework for the analysis of the relationships between environment and health.

This study suggests that future interventions to control weight gain by addressing the local social and physical environment may be promising. Indeed, obesity and excess weight can be reduced by interventions designed to promote a better, healthy environment. The results achieved provide evidence about the potential health effects of policies targeting different environmental features:

1. Changes to city form and urban design. Improving the availability of cycle-lanes and pedestrian streets, public spaces, parks and green areas, the connectivity of streets, and so on, will promote physical activity and social interaction.
2. Changes in land use. Diversity and proximity resulting from a vibrant mix of housing, shops and public facilities (compactness) should be promoted, in order to achieve forms of settlement that are denser and more compact, where facilities, ranging from public transport to shops, are nearby.
3. Reducing the burden of socioeconomic disadvantage and inequality, and levels of absolute and relative deprivation. These policies should be highlighted, since their effects could be decisive in different areas of life. Indeed, reducing inequality could improve both the access to material resources needed to lead a healthy life and the quality and strengths of social relationships, sources of social capital.
4. Targeting the indoor environment. While, on the one hand, the indoor environment can provide some measure of protection against outdoor hazards, on the other hand, housing that lacks basic facilities or is in poor physical condition could itself pose a risk to health. It should be stressed that

housing showed the strongest influence on the BMI of residents in the Lisbon Metropolitan Area, and so the potential health benefits related to this features are enormous.

5. Changing individual behaviour in order to reduce vehicle dependency and promote physical activity. Behavioural change will be encouraged by the changes listed above, which will decrease the need for motorised transport.

Issues related to urban sprawl, socio-economic inequalities, means of transport and social organization should be addressed in an integrated way, since they are a primary condition for the development of collective identities and feelings of belonging, necessary for sustainable communities. Sprawling, unsafe and pedestrian-unfriendly cities make people fat and unhealthy. But these conditions could be reversed, making places safe, pleasant and healthy for everybody: encouraging walking and cycling for short journeys; giving pedestrians priority (two-way pavements and cycle lanes, limited traffic, road and traffic-signs), instead of the omnipresent cars that characterize our cities (shaping a perception of fast lines as fat lines); improving the availability of sports facilities, including green areas and parks; creating "walkable" distances between facilities; improving the local social environment, strengthening identities, developing a sense of place and reducing levels of material deprivation and social inequality. These changes will encourage behaviour changes, such as increased cycling and walking. The time has come to reduce car dependency, increase physical activity (walking and cycling for recreation and transportation), and bring about a shift to communities that have been intentionally designed to facilitate physical (and mental) health. As David Rudlin and Nicholas Falk (1999, 147, 150) state:

> the way that we plan human settlement has an important role to play on increasing the sustainability of human activities … . The dense "walkable" city may therefore be the most environmentally efficient form of settlement for the majority of the population.

Area and community interventions, targeting the localization and availability of resources, housing adequacy, socioeconomic disadvantage and social capital, need to be designed and implemented. Individuals and neighbourhoods should both be targeted in order to enhance people's health. The local environment plays a major role in reducing the need and use of motorised transport, which has three main health benefits, direct or indirectly: 1. reduction of accidents and of noise and air pollution; 2. promoting regular physical exercise; 3. increasing social interaction and sense of place, since "social isolation and lack of community interaction are strongly associated with poorer health" (Wilkinson and Marmot 2002).

5.1. Study Uncertainties and Futures

Despite these results, it should be pointed out that no support was found for consistent relationships between BMI and some environmental features. Inconclusive results were achieved with bonding social capital/BMI, crime/BMI, and public transport accessibility/automobile dependence/BMI associations, recently proved elsewhere (Cohen et al. 2006; Kim et al. 2006; Poortinga 2006; Santana et al. 2006).

Perhaps, as Mohan et al. (2005) suggest, the inconclusive results were related with problems of collinearity between some environmental measures and the consequent impossibility of unpacking their relative effects in a combined model. Another possible explanation is related to the geographical level of data collection and analysis. Indeed, this study made use mostly of parish-level or ward-level measures that could distort real relationships, since these units are administrative areas, not true communities. In relation to crime, another problem could result from data inadequacy, since our data measures the number of reported crimes, as opposed to the fear and perception of crime, which has a well established relationship with health (Young et al. 2004).

Regarding social resources, our work seems to offer some support to a body of research into the influence of social environment on overall health and on BMI. Some of this research stresses the effects of social capital on health (Poortinga 2006). According to this perspective, health improvement could be achieved through investments not only in physical and material resources, but also through investment in social resources. In our work, the influence of material and immaterial resources on BMI was highlighted, although one of the social capital measures hasn't showed a significant influence on overweight BMI. We must return to this issue, putting forward the hypothesis that the bonding social capital indicator may not be an accurate representation of the neighbourhood social cohesion, generally identified as the relational dimension of social capital. It may also be that social bonds may be as conducive to poor health as good health behaviours.

However, the influence of social environment on BMI can be underlined, underlying the significance shown by political participation (linking social capital) and the potential to use the area disadvantage index to measure relative deprivation and socioeconomic inequalities. In relation to deprivation, it has been argued that social and material inequalities erode social capital by widening the gap between rich and poor, powerful and excluded (Kawachi 2000). McCulloch (2003) argues that socially disadvantaged neighbourhoods are more likely to lack environments conducive to the development of social organisation and social capital. Deprivation acts as a force that hinders mutual trust and the shared willingness to taking effective collective actions. Daily experiences with distrust, uncertainty and economic dependence are likely to weaker interventions for the common good, even when personal ties are strong. Thus, a deprivation index could be understood as a proxy of different levels of social organization.

Perhaps the main conclusion that we can draw from this debate is that our work about environmental effects on weight (and overall health) is just beginning and that more research is required into the contextual determinants of health: what they are; what they do; and how they can be measured. Moreover, in future studies, data and methodologies need to be improved in order to avoid possible bias related with them.

References

Andrews, G. and Kearns, R. (2005), 'Everyday health histories and the making of place: The case of an English coastal town', *Social Science and Medicine* 60: 2697–2713.

Backett-Milburn, K., Wills, W., Gregory, S. and Lawton, J. (2006), 'Making Sense of Eating, Weight and Risk in the Early Teenage Years: Views and Concerns of Parents in Poorer Socio-Economic Circumstances', *Social Science and Medicine* 63(3): 624–635.

Cohen, D., Finch, B., Bower, A. and Sastry, N. (2006), 'Collective Efficacy and Obesity: The Potential of Social Factors on Health', *Social Science and Medicine* 62: 769–778.

Cummins, S., Macintyre, S., Davidson, S. and Ellaway, A. (2005), 'Measuring Neighbourhood Social and Material Context: Generation and Interpretation of Ecological Data from Routine and Non-Routine Sources', *Health and Place* 11(3): 249–260.

Davies, A. (ed.) (2002), *A Physically Active Life through Everyday Transport. With a Special Focus on Children and Older People and Example and Approaches from Europe* (World Health Organization: Regional Office for Europe).

Day, R. (2007), 'Place and the experience of air quality', *Health and Place* 13: 249–260.

Diez-Roux, A., Merkin, S., Arnett, D., Chambless, L., Massing, M., Nieto, J., Sorlie, P., Szklo, M., Tyroler, H. and Watson, L. (2001), 'Neighborhood of Residence and Incidence of Coronary Disease', *New England Journal of Medicine* 345: 99–136.

Frank, L. and Engelke, P. (2001), 'The built environment and human activity patterns: exploring the impacts of urban form on public health', *Journal of Planning Literature* 16(2) 202–218.

Fullilove, M. and Fullilove, R. (2000), 'What's housing got to do with it?', *American Journal of Public Health* 90: 183–184.

Gikes, K. Ellaway, A. Santana, P. (2005), 'Obtaining area-level data in four countries: examples for crime, transport and leisure and recreation facilities', *Fourth Annual Conference of the International Society of Behavioral Nutrition and Physical Activity* (ISBNPA), Amsterdam: 37.

Jackson, R. (2002), 'Creating a healthy environment: The impact of the built environment on public health', in S. Srinnivasan, L. O'Fallon, and A. Dearry (eds), *Built Environment – Healthy Communities, Healthy Homes, Healthy People. Final Report.*

Jochelson, K. (2004), *The Public Health Impact of Cities and Urban Planning Report* (London Development Agency: King's Fund).

Kawachi, I. (2000), 'Income inequality and health', in L.F. Berkman and I. Kawachi (eds), *Social Epidemiology* (New York: Oxford University Press).

Kawachi, I. and Berkman, L. (2000), 'Social Cohesion, Social Capital, and Health', in L.F. Berkman and I. Kawachi (eds), *Social Epidemiology* (New York: Oxford University Press).

Kim, D., Subramanian, S., Gortmaker, S. and Kawachi, I. (2006), 'US State-and County-Level Social Capital in Relation to Obesity and Physical Inactivity: A Multilevel, Multivariable Analysis', *Social Science and Medicine* (forthcoming).

Leslie, E., Coffee, N., Frank, L., Owen, N., Bauman, A. and Hugo, G. (2006), 'Walkability of Local Communities: Using Geographical Information Systems to Objectively Assess Relevant Environmental Attributes', in *Health and Place* (forthcoming).

Lindstrom, M., Moghaddassi, M. and Merlo, J. (2003), 'Social Capital and Leisure Time Physical Activity: A Population Based Multilevel Analysis in Malmo, Sweden', *Journal of Epidemiology and Community Health* 57(1): 23–28.

Lochner, K., Kawachi, I., Brennan, R. and Buka, S. (2003), 'Social Capital and Neighborhood Mortality Rates in Chicago', *Social Science and Medicine* 56(8): 1797–1805.

Lopez-Zetina, J., Lee, H. and Friis, R. (2006), 'The Link Between Obesity and the Built Environment. Evidence from an Ecological Analysis of Obesity and Vehicle Miles of Travel in California', in *Health and Place* (forthcoming).

Macintyre, S., Ellaway, A. and Cummins, S. (2002), 'Place Effects on Health: How Can We Conceptualise, Operationalise and Measure Them?', *Social Science and Medicine* 55(1): 125–139.

Macintyre, S. and Ellaway, A. (1999), 'Local Opportunities Structures, Social Capital and Social Inequalities in Health: What Can Central and Local Government Do?', *Health Promotion Journal of Australia* 9(3): 163–170.

McCulloch, A. (2003), 'An Examination of Social Capital and Social Disorganisation in Neighbourhoods in the British Household Panel Study', *Social Science and Medicine* 5: 1425–1438.

McNeill, L., Kreuter, M. and Subramanian, S. (2006), 'Social Environment and Physical Activity: A Review of Concepts and Evidence', *Social Science and Medicine* (forthcoming).

Mohan, J., Twigg, L., Barnard, S. and Jones, K. (2005), 'Social Capital, Geography, and Health: A Small-area Analysis for England', *Social Science and Medicine* 60: 1267–1283.

Nogueira, H. (2006), 'Os Lugares e a Saúde. Uma abordagem da Geografia às variações em saúde na Área Metropolitana de Lisboa', PhD Thesis, University of Coimbra.

Nogueira, H. and Santana, P. (2005), 'Geographies of health and deprivation: relationship between them', in C. Palagiano and G. Santis (eds) *Atti dell' VIII Seminario Internazionale do Geografia Medica* (Roma: Perugia, Edizioni Rux)

Pearce, N. and Davey-Smith, G. (2003), 'Is Social Capital the Key to Inequalities in Health?', *American Journal of Public Health* 93: 122–129.

Pendola, R. and Gen, S. (2006), 'BMI, Auto Use, and the Urban Environment in San Francisco', in *Health and Place* (forthcoming).

Poortinga, W. (2006), 'Perceptions of the Environment, Physical Activity, and Obesity', *Social Science and Medicine* 63: 2835–2846.

Robert, S.A. and Reither, E.N. (2004), 'A multilevel analysis of race, community disadvantage, and body mass index among adults in the US', *Social Science and Medicine* 59: 2421–2334.

Rudlin, D. and Falk, N. (1999), *Building the 21st Century Home* (Oxford: Architectural Press).

Santana, P., Nogueira, H. and Santos, R. (2006), 'Impacts of Urbanisation on Weight Gain and Obesity in Portugal', in R. González (ed.), *Urban Changes in Different Scales: Systems and Structures* (Universidade de Santiago de Compostela, Publicacións).

Szreter, S. and Woolcock, M. (2004), 'Health by Association? Social Capital, Social Theory and the Political Economy of Public health', *International Journal of Epidemiology* 33(4): 650–667.

Sundquist, K. and Yang, M. (2006), 'Linking Social Capital and Self-Rated Health: A Multilevel Analysis of 11,175 Men and Women in Sweden', *Health and Place* (forthcoming).

Van Lenthe, F., Brug, J. and Mackenbach, J. (2005), 'Neighbourhood Inequalities in Physical Inactivity: The Role of Neighbourhood Attractiveness, Proximity to Local Facilities and Safety in the Netherlands', *Social Science and Medicine* 60: 763–775.

Veenstra, G., Luginaah, I., Wakefield, S., Birch, S., Eyles, J. and Elliot, S. (2005), 'Who You Know, Where You Live: Social Capital, Neighbourhood and Health', *Social Science and Medicine* 60 (12): 2799–2818.

Wilkinson, R. and Marmot, W. (eds) (2002), *Social Determinants of Health. The Solid Facts* (Copenhagen: World Health Organization).

Young, A., Russel, A. and Powers, J. (2004), 'The Sense of Belonging to a Neighbourhood: Can It Be Measured and Is It Related to Health and Well Being in Older Women?', *Social Science and Medicine* 59: 2627–2637.

Acknowledgements

The research for this chapter was financed by grant POCTI/GEO/45730/2002 from the Portuguese Foundation for Science and Technology (FCT) "Healthy Urban Planning". The authors would like to thank Pedro Pitta Barros, Rita Santos and Cláudia Costa.

Chapter 12

Sense of Place, Quality of Life and (g)local Struggles for Environmental Justice

Michael Buzzelli

1. Introduction

Sense of place, human territoriality and therapeutic landscapes are but a few of the concepts developed by humanists to understand the dynamic and reciprocal relationship between people and their environments, particularly the emotive connections. As this literature continues to grow a new lens on sense of place is environmental justice (EJ). Spurred by civil and minority rights movements that began in the 1960s in the United States, communities have mobilised to protect their homes and neighbourhoods from the inimical (perceived or real) effects of "environmental bads" such as waste incinerators and hazardous landfills. Mobilisation is inherently tied to sense of place and is expressed as territorial defence within EJ as communities who face threats to the quality of their environments and their health mobilise. To understand this new lens on sense of place this chapter will first outline the history of EJ mobilisation. Two case studies, one of EJ in sustainable planning in San Francisco and the other of grassroots mobilising in New Mexico, are then used to illustrate the connections between sense of place and EJ activism. But first we overview the conceptual basis of environmental justice, particularly as it flowered in the United States in the 1970s and 1980s.

2. Key Concepts in Environmental Justice

The EJ literature is nearly as old as the movement itself and can be read as a chronicle of community struggles to defend territory from external threats. In EJ that threat is invariably to environmental quality and its impact on human health. As a territorial defence the struggle is an expression of power, particularly by communities of low resistance who seek to both claim that "their territories" are targeted because of their marginalised identities (and thereby marginal territorial status) and that accordingly they shoulder an unfair burden of environmental degradation.

Some attribute the spark of EJ to the Love Canal (Niagara Falls, New York) disaster of 1978 when, after years of toxic waste dumping, heavy precipitation leached chemicals into soils and nearby homes and forced emergency measures including evacuation (Fletcher 2003). Joining up with a civil rights movement already in full swing, Love Canal raised public awareness and galvanized the anti-toxics movement, particularly among low status and minority communities. The movement gained momentum when, in 1982, rural Warren County (North Carolina), primarily a poor African American community, was confirmed as the site for a toxic dump to store soil that had been illegally contaminated with PCBs (Goldman 1996). Reaction was immediate and the images all too familiar: protests were followed by law suits but it still took until 2004 for the site to be decontaminated.

Alongside community mobilising that inevitably forms part of these cases, what these cases also did was activate a substantial academic literature that is at once mature and still flourishing today. Year 2007 marks the twentieth anniversary of a landmark study, *The Commission for Racial Justice*, published by the United Church of Christ in 1987 (UCC 1987).[1] The US General Accountability Office published its own explosive study in 1983, *Siting of hazardous waste landfills*, and shortly following the UCC study came one of the most famous titles in the EJ literature, *Dumping in Dixie*, by Robert Bullard (1990). These were heady days and the literature has since mushroomed. Accordingly, the *Environmental Justice Resource Centre* at Clark Atlanta University commissioned a twentieth anniversary follow-up to the UCC study titled *New Toxic Wastes and Race at Twenty* (Bullard et al. 2007; see also http://www.ejrc.cau.edu/TWARTFinal.htm). Updating the *Commission* with year 2000 US census data, pollutant data and geographic information systems (GIS), the study finds that race – more than income and property values for example – remains the most important predictor of the location of hazardous waste facilities in the United States. From the original *Commission*, we read that "people of colour were twice as likely as Whites to live in a community with a commercial hazardous waste facility and three times as likely to have multiple facilities." As we shall see later the methodological basis of such conclusions is regarded by some as weak but these kinds of summary statements conclusions are politically very powerful.

The EJ literature is now well developed in the US and emergent elsewhere and reflects grassroots concern over, and subsequent study of, a range of health hazards including air pollution, toxic waste sites, toxic spills, noxious facility allocation, waste incineration, pesticides, lead and heavy metals exposure, just to name a few. Notwithstanding methodological questions, there was an important policy response to these emerging insights at the highest level in the United States. President Clinton issued Executive Order 12898: "Federal actions to address environmental justice in minority populations and low-income populations" (Pres. WJ Clinton, Feb. 1994). All of this activity reflects the unequal exposure of environmental (technological) health hazards among communities and areas defined as predominantly poor and visible minorities. What is meant by "unequal" here is that exposures, or potential exposures, are more often found in marginalized communities. Thus we arrive at questions of fairness and equity. At the root of the EJ movement is the belief that poor and minority communities bear a disproportionate burden of hazards. The reason

1 Environmental racism, a term coined by Dr. Benjamin Chavais, lead author of the UCC Commission, is equivalent to environmental justice but prioritises the role of race and racism in conditioning community exposure to hazards.

unequal exposures are seen as disproportionate is that these communities neither generate the "regrettable necessities" of the production and consumption sectors nor benefit from the living standards and quality of life that follow. And yet they usually bear the lion's share of exposure to the air pollution and ground water contamination (and potential adverse health effects) that flow from these development processes.

Relatedly the term "justice" itself signifies a particular concern with the health effects of exposure to environmental health hazards. The intent here is to draw a simple dichotomy between justice and health effects though it bears noting the profound political philosophy that lies behind notions of justice (Fainstein 2000; Smith 1993; Young 1990) that often informs the EJ literature (e.g. Holifield 1991). Environmental economics, for example, has a long-standing interest in environmental "equity" although health is only one of several concerns over environmental disbenefits. Instead the term "justice" reflects a concern over the (potential) health effects of technological hazards. Thus questions of norms and fairness signal the politically charged – and perhaps politicised – nature of health within the EJ discourse. Burger (1990) noted this in arguing that to some extent environmentalism is fuelled by the perceived or real impacts of environmental change on human health rather than environmental stewardship *per se*. Indeed this may go to explaining, as detailed further below, why otherwise scientifically "weak" study designs in early EJ research could nonetheless be used for claims-making and policy development in the United States. Nonetheless, for some what makes the EJ movement unique is the novel focus amongst civil rights advocates over questions of environment *and* health alongside issues of economic and social justice (Taylor 2000). At the same time, the EJ movement and research stream addresses questions of distribution more than absolute reductions of environmental toxicants (as in Industrial Ecology). EJ is primarily a literature concerned with horizontal distributive justice – where toxicants are distributed and who is exposed – rather than emissions reduction in the first instance.

Another important issue within EJ can be summarised as a "chicken and egg" conundrum concerning the timing of technological hazards and low status communities. On the one hand technological hazards may concentrate disproportionately in existing communities of least resistance or, on the other, hazards suppress land values making properties affordable to those of lower status. In this respect Been (1993) asked "what's fairness got to do with", arguing that conclusions like that of the UCC study cited earlier reflect a natural evolution of the land market: that Locally Unwanted Land Uses (LULUs) create affordable accommodation in proximate real estate. In contrast Pulido (1996) has made the point that wider structural forces, such as racism, account for both low status communities with technological hazards and that from this point of view residential "choice" is an abstraction. The debate goes to and fro though some have provided analytical answers. For example, Hamilton (1995) demonstrates econometrically that existing commercial hazardous waste facility expansion across the USA is closely associated with communities composed of racial minorities.

Finally one issue lurking in what has been said so far – about communities of least resistance, racial minorities, the poor – is the "race or class question". EJ researchers have spent some time debating which sets of socioeconomic markers are more important in correlating with hazards. In *Faces of Environmental Racism*,

Westra and Lawson (2001) ask the “race or class” question to begin their overview of the literature and argue that race is the more important identity marker (for similar conclusions see Bryant and Mohai 1993; Bullard 1993; Hofrichter 1993). Others ask how it is possible to separate low social status from race in the US context. Others still argue that race has been used all too loosely in EJ research without taking account of the racialisation process (Pulido 2000).

These then are the salient themes within EJ research. They also speak to the connection with sense of place. For instance it matters little whether we can answer the “race or class” question from the point of view of understanding the role of place identity in generating environmental activism. The same can be said of resolving whether community targeting or marginal community composition came first. The point is that we arrive, inevitably some would argue, at an unjust distribution of environmental bads that spurs community mobilisation. To be sure the EJ movement can be criticised for its focus on pollution distribution rather than a wider environmental concern with pollution reduction but the standpoint of affected communities is understandable: we could hardly expect marginalised communities which feel targeted to organise around pollution reduction when they anticipate to continue carrying more than their share of toxicants. Thus feeling their communities are targeted because of their marginal identity and status results in a particular kind of sense of place, one that is defined by territorial defence aimed at power and control over the quality of local environments. The case studies presented below, while different in origin and focus, speak to questions of power and identity in environmental decision making and place-definition.

3. EJ at the Grassroots

So what does it mean to be exposed to environmental health hazards? What are communities concerned about and how do they respond? To answer these questions we draw from recent examples of research that demonstrate, using a range of data types and methods, exposure to technological hazards. We then turn to two case studies that illustrate the relationship between sense of place and policy development at the grassroots.

In 1995 Susan Cutter characterised the EJ literature as having two largely exclusive streams. The first, process studies, critically trace the historical development of environmental injustice. Perhaps the best know example is in Pulido’s (2000) research on Los Angeles. Aimed at exposing the spatiality of racism and how it underpins environmental racism, Pulido demonstrates how “white privilege” has led to simultaneous decentralisation and suburbanisation of the white population in Los Angeles. Over the twentieth century, she argues, whites have secured relatively cleaner environments while facility siting processes – the usual subject of EJ research – have at the same time targeted communities of least resistance within the region. The end result, shown in several maps and tables, is significant spatial coincidence of racial minority communities and uncontrolled toxic waste sites, Toxic Release Inventory (TRI) facilities and Transfer, Storage and Disposal Facilities (TSDFs).

Outcome studies, by contrast, examine the (usually spatial) correlation between environmental health hazards and communities "cross-sectionally", or at a point in time. For example, Buzzelli and Jerrett (2004) analysed the neighbourhood-level exposure to air pollution (annual average total suspended particles, TSP) in Hamilton, Canada in 1996. They found that low education and low property values correlated significantly with air pollution as did visible minority markers of neighbourhood status, namely Latin-American (positively associated with air pollution) and South-Asian (negative) groups. On top of the substantive conclusions this study also highlights the complex relationships between alternative measures of community socioeconomic status.

These brief examples illustrate that EJ is a research stream concerned with environmental health hazard exposures. But the struggle for environmental justice is just that – a struggle – for it pits priorities of employment, standard of living and development against environmental quality and human health. At the same time, local movements confront processes that often reach beyond the boundaries of the local such as capital mobility, foreign direct investment and unpriced environmental degradation. In essence the struggle for environmental justice is that of quality of life against standard of living and of the local against the global. How do these priorities and tensions draw upon and reinforce a community's sense of place?

3.1 Case 1: San Francisco's Sustainable City Plan

In 1993 a wide range of stakeholder communities in the San Francisco Bay area initiated a process to develop sustainable planning principles for the region (http://www.sustainable-city.org/index.htm). This included over 400 volunteers and thousands of hours of consultations with environmental NGOs, private business, local universities and of course the City itself. Modelled after the European Community's Agenda 21 Implementation Plan (UN Strategy for sustainable development) the goals and objectives of *Sustainable City* became official policy of the City and County of San Francisco in 1997 alongside the City's new (1996–7, ongoing) Department of the Environment.

What is unique about this sustainability planning exercise is that EJ was incorporated as an explicit theme (unlike Agenda 21), making San Francisco one of the earliest municipalities in the United States to develop EJ principles in its planning process (Agyeman et al. 2003). Of course environmental topics such as air pollution (the others were biodiversity; energy, climate change and ozone depletion; food and agriculture; hazardous materials; human health; parks, open spaces and streetscapes; solid waste; transportation; water and wastewater) were part of the plan though EJ (along with economy and economic development; municipal expenditures; public information and education; risk management) was developed as a cross-cutting theme. Thus EJ was a particular lens on localised issues that provided some impetus for considering EJ in this planning process. For instance, in Bayview-Hunters Point district, residents who are primarily low-income African Americans were working with the City's Public Health Department to undertake an environmental assessment for a controversial proposed power plant. The district is already one of the most polluted in the region. Other localised issues relate to alternative levels of city service provision (such as waste collection) and lead-based paint exposures.

However the guiding principle for inclusion of EJ in *Sustainable City* was less reaction to localised issues but the need to incorporate ideals of public participation in planning, particularly among under-represented groups such as racial(ised) minorities. From *Sustainable City*:

> From the perspective of environmental justice activists, sustainability must include a process of collective decision-making and address issues of social inequality and racism as well as ecological degradation … One of the most important aspects of environmental justice is the question of participation. Any plan for the City's sustainability should reflect the views and perspectives of San Francisco's multi-racial, multi-ethnic communities and not only just those of people with the time to attend drafting meetings. For the Sustainable San Francisco project to serve as an effective planning tool, community outreach efforts must be undertaken and public hearings and planning sessions conducted beyond those conducted during the summer of 1996 [when the scoping process was initiated] (accessed 16 May 2007).

This line of thought persists with the City's Department of the Environment which, on its EJ web site, states "Environmental justice is the fair treatment and meaningful involvement of all people, regardless of race, ethnicity, income, or education level, in environmental decision-making". The crown jewel of the Department's EJ program is its grants-in-aid funding to communities to address pollution, energy and public health concerns (initiated in 2000, ongoing). Over nine million dollars have been granted to 25 different projects. Among the results, for example, is a new (2005) Bayview-Hunters Point Farmers' Market that provides low-income residents with regular local access to fresh produce. This and other programmes have received a generally positive review in an independent audit of the programme's operation and community impacts (TechLaw, Inc. 2006).

3.2 Case 2: Southwest Organizing Project

The SouthWest Organizing Project (http://www.swop.net/) is a multi-issue grassroots organisation established in 1980 in New Mexico. Its mission is to "empower communities to realize racial and gender equality and social and economic justice". SWOP's members and organisers come from the Chicano (indigenous Mexicans), Native American, Mexican immigrant and smaller African American and Asian/Pacific Islander populations of New Mexico.

Not unlike *Sustainable City* in San Francisco, SWOP takes a holistic view of environment and defines its scope of interest, in a mantra common to EJ discourse, as "justice where we live, justice where we work, justice where we play." Therefore also a cross-cutting theme, one sees EJ in all of SWOP's organising activities from inception to present. As chronicled on its web site: in 1984 SWOP initiated a voter registration drive in Albuquerque and has since registered over 30,000 citizens; in 1988 SWOP worked with community residents to develop its Community Environmental Bill of Rights which lays out basic principles for environmental justice. The Bill of Rights was endorsed by Congressman Steve Schiff, present NM

Attorney General Tom Udall and the Bernalillo County Commission; in 1990 SWOP hosted the Regional Activist Dialogue, which brought together activists from eight southwestern states. Out of the event, the Southwest Network for Environmental and Economic Justice was formed. Today the network consists of 70 organizations in the Southwest and northern states of Mexico fighting for environmental and economic justice; in 2004 SWOP helped plan the New Mexico Environment Department and EJ listening sessions were held in four communities. Input gathered from these sessions informed the development of an EJ legislative bill, called the New Mexico Healthy Communities Act. Several other examples are listed around such issues as drinking water quality, local economic development and citizen representation.

Implicit in these examples are the overlapping themes within the scope of SWOP's activities, stated on its web site as public/community participation, economic justice and of course EJ. It is stated:

> Development in New Mexico is heavily influenced by outside forces that exercise great political control in the State. Since the 1840s, extraction type industries – mining, cattle ranching, timber, nuclear weapons development and now "high tech" electronics – have exploited our natural, labour and economic resources … People of color have paid the greatest price for this history. New Mexico is one of the poorest states in the Union and people of color are the majority. (Accessed 17 May 2007)

Reference to "outside forces" includes investment from within the US as well as foreign direct investment. Here we see the g(local) nature of environmental justice organising.

How has SWOP aided communities to respond? In one ongoing project involving a global microchip manufacturer, SWOP held year-long meetings with the corporation, facilitated by New Mexico Senator Pro Tempore Manny Aragon, to learn about its planned expansion into the State and develop recommendations for the State Legislature. Since the expansion, however the relationship has not developed as SWOP would have hoped. This has prompted organisers to pursue other strategies such as showing residents how to build their own albeit "low-tech" but effective air pollution samplers to use when they experience acute symptoms of air pollution exposures such as eye, nose and throat irritation. The "Bucket Brigade" (because the makeshift instrument involves use of an ordinary bucket, or pail, see Figures 12.1 and 12.2) have reported chemicals exceeding regulatory standards and others that are not authorised for use at the Rio Rancho microchip factory (organising is centred in Rio Rancho and nearby Corrales). Despite the "low-tech" approach, reports of the samples have attracted considerable attention, so much in fact that via a private donation of $50,000, residents worked with SWOP to purchase and set up a sophisticated continuous air pollution monitor to directly measure emissions from the factory (results reported on Scorecard, an independent pollution information web site). Unlike *Sustainable City*, SWOP is an example of collective environmental activism that not only includes the grassroots but is built from the ground up, as it were. SWOP has also developed around activism in respect to specific issues that have emerged within New Mexico communities. Thus SWOP exemplifies EJ organising at the interface of (g)local struggles that so often pit development against environmental (health) concerns.

Figures 12.1 and 12.2 The "Bucket Brigade"

Source: The SouthWest Organizing Project (http://www.swop.net/).

What both SWOP and *Sustainable City* demonstrate is, first, that citizen inclusion and participation is necessary to achieve EJ ideals. Second, EJ is an inter-sectoral movement. Sustainability in San Francisco's planning could not be achieved without due consideration of social and spatial inequities, a point also made by leading EJ researchers (e.g. Agyeman et al. 2003; Dobson 2003). Third, these two cases also reveal the uniqueness of the EJ movement: that it brings together stakoholders in a new kind of movement. According to Taylor (2000, 42), EJ "... is the first paradigm to link environment and race, class, gender and social justice concerns in an explicit framework". Thus identity, whether based on race, class or some other maker, is central to place identity in EJ planning and organising.

Taking this further, what is both implicit and yet obvious about these cases and practically any localised EJ issue is that place identity is infused into the territorial defence of communities who feel targeted in environmental decision-making (Hayden 1995; Sack 1986) that underlies the protection of the local environment. Once again drawing from SWOP (accessed 26 September 2007), the Community Environmental Bill of Rights developed with organising partners over years of activism outlines eight categories of "rights": (1) to clean industry; (2) to be safe from harmful exposure; (3) to prevention; (4) to know [of developments in NM]; (5) to participate; (6) to protection and enforcement; (7) to compensation; (8) and to clean up. In each of these we can tease out degrees of sense of place and territoriality. Among the most obvious are: "industry that will contribute to the economic development of our communities and will enhance the environment and beauty of our landscape ... the right to participate in forming public policy that prevents toxic pollution from entering our communities" and in a novel and progressive geographic expression "We have the right to be insured that our problem is not transferred to other communities".

These principles raise a host of questions about the ways in which we assess hazards, exposure and risk to populations as well as the origins of these notions from within particular locales and residents' own sense of place. We turn next to some of the methodological issues implicated in these statements and return in the concluding discussion to wider issues of sense of place for EJ organising, public participation, advocacy, planning and the like.

4. Gaining Momentum, Addressing Gaps

EJ research is growing, spreading abroad as the quality of research in the US also rises to address some important gaps in the literature (Pellow 2006). Meanwhile the policy agenda in the US is also filling in and we might argue – at the risk of scorn from those wishing more policy development – that EJ has been a more successful movement and policy arena than a research literature. In any event we can take all of this development as a sign of maturity amidst continued growth.

One gap concerns health effects of environmental hazards. Even though health impacts are one of the principal rationales for EJ research and activism, literature searches show that remarkably few studies have actually undertaken health effects testing in the epidemiologic sense of risk assessment (Buzzelli 2007, 5). This has begun to change only recently (Rogers and Dunlop 2006; Finkelstein et al. 2005; Wheeler and Ben-Shlomo 2005; Harner et al. 2002). For example, studies in southern California (Morello-Frosch et al. 2001; Pastor et al. 2005) have used emissions inventories (including NATA) to assess the health impacts of a range of sources, reporting mobile/transportation sources as most important for lifetime cancer risk, but especially for racial minorities. These results, they argue, point to the need for land use and public policy development on transportation emissions to reach beyond the typically targeted large-facility emissions. More broadly, studies like these are beginning to undergird the much more established literature that advocates environmental health remediation but is devoid of evidence of health effects.

A similar issue concerns the environmental epidemiology and exposure assessment that is almost entirely absent in the EJ literature owing to the lack of reliable monitoring data that is needed to build such evidence. As suggested by the case of SWOP, communities do not usually have the resources for their own monitoring and regulatory monitoring can rarely be used to the standard needed for human exposure assessment. Thus while we are aware of the presence of environmental hazards we know rather less about the health risks posed by exposure. Here again we see that the minority rights and civil liberties basis of EJ may veil otherwise questionable research quality for the sake of claims-making. In an incisive review of EJ research Bowen (2002) classifies the several studies into high-, medium- and low-quality categories and argues that anything but high quality research can be harmful for policy and management decisions. He goes on to say [with respect to medium-quality research] that "the flaws and limitations were substantial enough to judge that the research was not of sufficient quality, say, to recommend using it in court of law or in an actual policy or administrative decision related to environmental justice" (Bowen 2002, 7).

Relatedly, the methods used to assign exposure and assess relationships are often regarded as inadequate. For instance Anderton et al. (1994) fired an important early salvo by demonstrating that the strong spatial association of minority communities and toxic waste sites found in an early flagship *Dumping in Dixie* (Bullard 1990), changed at different spatial scale of analysis. Seen in this light it is not surprising that health studies are far fewer than studies of the spatial coincidence of hazards and low status communities and, more broadly, why some argue that the literature has produced only scant evidence that meets data quality and methodological standards sufficient for proper health research or litigation (Brulle and Pellow 2006; Downey 2006; Mohai and Saha 2006; Bowen 2001).

Notwithstanding these shortcomings that, to be sure, are being addressed in recent work, the EJ research agenda is spreading. And it's no wonder: in the US the explosiveness of the issue, coupled with early foundational studies noted earlier, led to ample policy development. Executive Order 12898 noted earlier was just one step. The prior year, in 1993, the US Environmental Protection Agency had already developed its own EJ programme within its Compliance and Enforcement division. Its EJ programme incorporates the National Environmental Justice Advisory Council (NEJAC), granting programmes and an internship programme. Development at the State and local level, like Sustainable City (San Francisco), meanwhile, is uneven but also proceeding (Warner 2002). Seeing this, researchers have begun to ask questions about EJ in other countries including in Australia, Canada, the United Kingdom and elsewhere (Buzzelli et al. 2006; Lloyd-Smith and Bell 2003; McCleod et al. 2000; Brainard et al. 2002; Mitchell and Dorling 2003).

As the geographical scope of EJ interests and policy continue to grow, the literature is beginning to show signs of maturity. First, whereas early studies – again primarily in the US – seemed to monotonically repeat the received wisdom of *de facto* environmental injustice/racism, we are now beginning to see studies that also furnish subtle and sometimes contrarian results. The Anderton et al. (1994) revision cited above is one example. In another, Most et al. (2004) address the difficulty of population assignment for devising comparison groups and assigning exposure to transportation externalities (noise in their study). They demonstrate that alternative spatial scales can significantly alter study results (see also Brulle and Pellow 2006; Downey 2006; Ringquist 2005). Second and relatedly, we are beginning to see an ever more diverse range of issues in addition to the air pollution and toxic waste sites that have dominated the discourse. We now may ask what aspects of the built environment in general, such as the walkability of neighbourhoods and the opportunity to buy fresh produce, are part of the environmental goods and bads that produce environmental injustice.

5. Discussion

Following the case studies and this expanding scope of inquiry, we are left with the question of sense of place: what role does it play in informing our insights on environmental justice? On impelling EJ organising in particular places? On what we ought to consider important for inclusion in EJ-informed policy and planning?

The example of SWOP is useful for projecting some of these issues. One can envision environmental rights, as expressed in the Community Environmental Bill of Rights, as growing out of citizens' rootedness and in particular their territorial defence of everyday spaces. The Bucket Brigade is an outward expression of these emotive connections with Rio Rancho and surroundings. Indeed there is a critical point here: the very industrial emitters maligned in SWOP organising surely represent a lion's share of jobs, investment and development in the region; an island of investment in a sea of under-development. In spite of this, SWOP organisers, as represented in their expressed environmental rights, would forego – or certainly significantly modify – that standard of living in defence of quality of life. They adopt a longer-term (rooted) view of development in Rio Rancho and NM more broadly.

For residents of Rio Ranch, and those of any place with charged EJ issues, environmental hazards are a dystopian reality. Major roadways, industrial facilities and waste incinerators are constant reminders of the unhealthfulness of their surroundings. What is missed in virtually all EJ research are the impacts of this day-to-day grinding. The effect of environmental health hazards can go beyond narrowly defined dose-response relationships, even (or perhaps despite) when experts fail to appreciate the psychosocial and emotional. For example in Camelford UK, spillage of aluminium sulphate into the drinking water supply in 1988 was deemed by experts to have no long-term impact on a range of cognitive functions. Ongoing complaints by locals was attributed to sensitisation bias. However a follow-up study one decade later opened up the possibility that lay perceptions of longer-term effects warranted at least further study (e.g. Altmann et al. 1999). In another example, a multimethod study of landfill siting in Southern Ontario found that residents could be more concerned with their lack of inclusion in the planning process than with more direct environment and health linkages (c.f. Elliott et al. 1993; Eyles et al. 1993).

As these examples suggest a major blind-spot in EJ research is analysis of the psychosocial impacts of environmental hazards quite apart from physical exposure itself. At the very least we can add this to the growing agenda of health effects research within EJ. But the implications are broader still: In EJ territoriality necessarily grows out of lay perceptions of the healthfulness (or otherwise) of places that must also be included in advocacy and planning. Brown's incursions (1987, 1997) on "popular epidemiology" speak to the point: In Woburn in the 1970s and 1980s residents organised around suspected linkages between leukaemia clusters and toxic waste disposal in what has become one of the more storied environmental activism cases in the United States. Through this case Brown argues for formal recognition of popular epidemiology in which "citizen scientists" covey their understanding of risk to authorities and experts alongside political mobilisation. What are we missing if we ignore the role of sense of place in EJ organising? Easier exchange of information? Does meaningful inclusion allow citizens to help set the terms of reference for health impacts and decision-making? One can imagine so: after all that is precisely what is demanded in the Community Environmental Bill of Rights and underpins San Francisco's *Sustainable City* plan. Although Woburn does not figure in the annals of EJ research it clearly highlights a new dimension of inquiry where socioeconomic and racial disparities are front and centre in environmental health disparities.

These questions represent opportunities for EJ research and organising. Meanwhile the movement itself is growing ever more complex. For example EJ is becoming entwined with the environmental lens on energy and climate change. Whereas nuclear energy was once an obvious motivation for environmentalists it now seems a meaningful alternative – at least for some – to hydrocarbon energy production. For example, a Nuclear Energy Institute (NEI 2005) study found 83 per cent of Americans living near to an existing nuclear power plant favour nuclear energy and 76 per cent would accept construction of a new plant nearby. Although the Institute is a trade organisation (and seen by some as a lobby), we are still led to ask: How will new priorities and perspectives like these play out within the EJ movement? In this example, will the once easy alliance against nuclear power now create fissures within the EJ movement to contest space and place amongst previously allied agents?

As the answer to this question unfolds it will be interesting to trace EJ mobilising and research across national boundaries to learn how the movement manifests in alternative contexts. In the Canadian context, for example, how will First Nations land claims – an issue over which Canada continues to receive scorn from the international community – impel or redefine EJ as we know it? In Europe, on the other hand, one can envision that European Union environmental regulations will be seen as impositions on member states as common standards of emissions, remediation, public health and public participation grow. But these are to some extent imponderables left to history to record and interpret. What we do know for certain is that communities will continue to be sensitised to the quality of their local environments and their (potential) health impacts. Coupled with questions of fairness and proportionality, the EJ movement and research agenda is not only firmly established but certain to grow further still.

References

Agyeman, J., Bullard, R. and Evans, B. (2003), 'Introduction: Joined-up thinking: Bringing together sustainability, environmental justice and equity' in J. Agyeman, R. Bullard and B. Evans (eds), *Just Sustainabilities: Development in an Unequal World* (Cambridge: MIT Press).

Anderton, D., Anderson, J., Oakes, J. and Fraser, M. (1994), 'Environmental equity: The demographics of dumping', *Demography* 31: 229–48.

Been, V. (1993), 'What's fairness go to do with it? Environmental justice and the siting of locally undesirable land uses', *Cornell Law Review* 78: 1001–36.

Bowen, W. (2001), *Environmental Justice Through Research-based Decision-making* (New York: Garland).

—— (2002), 'An analytical review of environmental justice research: What do we really know?', *Environmental Management* 29: 3–15.

Brainaird, J., Jones, A. and Bateman, I., et al. (2002), 'Modelling environmental equity: access to air quality in Birmingham, England', *Environment and Planning A* 34: 695–716.

Brown, P. (1987), 'Popular epidemiology. Community Response to Toxic Waste-Induced Disease in Woburn, Massachusetts', *Science, Technology and Human Values* 12: 3–4, 78–85.

—— (1997), 'Popular Epidemiology Revisited', *Current Sociology* 45 (3): 137–156.

Brulle, R. and Pellow, D. (2006), 'Environmental justice: Human health and environmental inequalities', *Annual Review of Public Health* 27: 103–24.

Bryant, B. and Mohai, P. (1993), *Race and the Incidence of Environmental Hazards* (Boulder: Westview Press).

Bullard, R. (1990), *Dumping in Dixie: Race, Class and Environmental Quality* (San Francisco: Westview Press).

Bullard, R., Mohai, P., Saha, R. and Wright, B. (2007), *New Toxic Wastes and Race at Twenty: Grassroots Struggles to Dismantle Environmental Racism in the United States* (New York: United Church of Christ).

Burger, E.J. (1990), 'Health as a Surrogate for the Environment', *Daedalus* 119(4), 133–153.

Buzzelli, M., Su, J., Le, N. and Bache, T. (2006), 'Health hazards and socio-economic status: a neighbourhood cohort approach, Vancouver, 1976–2001', *Canadian Geographer* 50: 376–91.

Buzzelli, M. (2007), 'Bourdieu does environmental justice? Probing the linkages between population health and air pollution epidemiology', *Health and Place* 13: 3–13.

Cutter, S. (1995), 'Race, class and environmental justice' *Progress in Human Geography* 19: 107–18.

Dobson, A. (2003), 'Social justice and environmental sustainability: Ne'er the twain shall meet?', in J. Agyeman, R. Bullard and B. Evans (eds) *Just Sustainabilities: Development in an Unequal World* (Cambridge: MIT Press).

Downey, L. (2006), 'Environmental inequality in metropolitan America in 2000', *Sociological Spectrum* 26: 21–41.

Elliott, S.J., Taylor, S.M., Walter, S., Stieb, D., Frank, J. and Eyles, J. (1993), 'Modelling psychosocial effects of exposure to solid waste facilities', *Social Science and Medicine* 37(6): 791–805.

Eyles, J., Taylor, S.M., Johnson, N. and Baxter, J. (1993), 'Worrying about waste: Living close to solid waste disposal facilities in southern Ontario', *Social Science and Medicine* 37(6): 805–812.

Fainstein, S. (2000), 'New directions in planning theory', *Urban Affairs Review* 35: 451–78.

Finklestein, M., Jerrett, M. and Sears, M. (2005), 'Environmental inequality and circulatory disease mortality gradients', *Journal of Epidemiol Community Health* 59(6): 481–487.

Fletcher, T.H. (2003), *From Love Canal to Environmental Justice: The Politics of Hazardous Waste on the Canada-US border Peterborough* (Peterborough: Broadview Press).

Goldman, B. (1996), 'What is the future of environmental justice?', *Antipode* 28: 122–41.

Hamilton, J. (1995), 'Testing for environmental racism: Prejudice, profits, political power?', *Journal of Policy Analysis and Management* 14: 107–32.

Harner, J., Warner, K., Pierce, J. and Huber, T. (2002), 'Urban environmental justice indices', *Professional Geographer* 54: 318–31.

Hayden, D. (1995), *The Power of Place: Urban Landscapes as Public History* (Cambridge: MIT Press).

Hofrichter, R. (1993), *Toxic Struggles: The Theory and Practice of Environmental Justice* (Philadelphia: New Society Publishers).

Holifield, R. (2001), 'Defining environmental justice and environmental racism', *Urban Geography* 22: 78–90.

Lloyd-Smith, M. and Bell, L. (2003), 'Toxic disputes and the rise of environmental justice in Australia', *International Journal of Occupational and Environmental Health* 9: 14–23.

McCleod, L., Jones, L., Stedman, A., Day, R., Lorenzi, I. and Bateman, I. (2000), 'The relationship between socio-economic indicators and air pollution in England and Wales: implications for environmental justice', *Regional Environmental Change* 1: 78–85.

Mitchell, G. and Dorling, D. (2003), 'An environmental justice analysis of British air quality', *Environment and Planning A* 35: 909–29.

Mohai, P. and Saha, R. (2006), 'Reassessing racial and socioeconomic disparities in environmental justice research', *Demography* 43: 389–399.

Morello-Frosch, R., Pastor, M. and Sadd, J. (2001), 'Environmental justice and Southern California's "riskscape": The distribution of air toxics exposures and health risks among diverse communities', *Urban Affairs Review* 36: 551–78.

Most, M., Sengupta, R. and Burgener, M. (2004), 'Spatial scale and population assignment choices in environmental justice analyses', *Professional Geographer* 56(4): 574–86.

NEI (Nuclear Energy Institute) (2005), *Questions for EPZ Survey: FINAL August 2005* . NEI and Bisconti Inc., 2610 WOODLEY PLACE NW, Washington DC.

Pastor, M., Morello-Frosch, R. and Sadd, J. (2005), 'The air is always cleaner on the other side: Race, space, and ambient air toxics exposures in California', *Journal of Urban Affairs* 27(2): 127–148.

Pellow, D.N. (2006), 'Transnational alliances and global politics: New geographies of urban environmental justice struggles', in N. Heynen, M. Kaika and E. Swyngedouw (eds), *In the Nature of Cities: Urban Political Ecology and the Politics of Urban Metabolism* (New York: Routledge).

President William Jefferson Clinton. Federal actions to address environmental justice in minority populations and low-income populations. 12989. 1994. Ref Type: Statute.

Pulido, L., Sidawi, S. and Vos, R. (1996), 'An archaeology of environmental racism in Los Angeles', *Urban Geography* 17: 419–39.

Pulido, L. (2000), 'Rethinking environmental racism: White privilege and urban development in southern California', *Annals of the Association of American Geographers* 90(1): 12–40.

Ringquist, E. (2005), 'Assessing evidence of environmental inequities: A meta-analysis', *Journal of Policy Analysis and Management* 24(2): 223–247.

Rogers, J. and Dunlop, A. (2006), 'Air pollution and very low birth weight infants: A target population?', *Pediatrics* 118(1): 156-164.

Sack, R.D. (1986), *Human Territoriality: Its Theory and History* (Cambridge: Cambridge University Press).

Smith, D. (1994), *Geography and Social Justice* (Oxford: Blackwell).

Taylor, D. (2000), 'The rise of the environmental justice paradigm: Injustice framing and the social construction of environmental discourses', *American Behavioral Scientist* 43(4): 508–580.

TechLaw I. (2006), *Independent Evaluation of the Environmental Justice Grant Program*. TechLaw, 90 New Montgomery Street Suite 1010 San Francisco, CA 94105.

UCC (United Church of Christ) CfRJ. (1987), *Toxic Wastes and Race in the United States* (New York: United Church of Christ, Commission for Racial Justice).

USGAO (United States General Accounting Office) (1983), *Siting of Hazardous Waste Landfills and their Correlation with Racial and Economic Status of Surrounding Communities* (Washington: Government Printing Office).

Warner, K. (2002), 'Linking local sustainability initiatives with environmental justice', *Local Environment* 7(1): 35–47.

Westra, L. and Lawson, B. (2001), *Faces of Environmental Racism: Confronting Issues of Global Justice* (Oxford: Rowman and Littlefield).

Wheeler, B. and Ben-Shlomo, Y. (2005), 'Environmental equity, air quality, socioeconomic status, and respiratory health: a linkage analysis of routine data from the Health Survey for England', *Journal of Epidemiol Community Health* 59: 948–954.

Young, I. (1990), *Justice and the Politics of Difference* (Princeton: Princeton University Press).

Chapter 13

In Search of the Place-identity Dividend: Using Heritage Landscapes to Create Place Identity

Gregory Ashworth

Few things are more important to a social group than its sense of belonging, not only to each other but to a place. What has sustained peoples in exile, from Babylon onwards, has been the possibility of one day returning home.

Lord Justice Sedley, Ruling in Court of Appeal, London, on the appeal of Chagos islanders right to return, 2007

The Assertions

This chapter poses a simple, and to some no doubt unnecessary, question, namely 'can heritage be used to create landscapes (and their urban manifestation, townscapes) in order to shape or enhance places identities, thereby increasing the well-being of those who experience them and, through that, contribute a recognisable and measurable dividend to the quality of life?' More specifically, 'can places be planned with the objective of revealing, preserving, enhancing or inventing local place identity and particularly can heritage be used as an instrument for achieving this objective?' The question, 'should this be done?' is usually implicit. To be clear on the relationships inherent in this question, landscape is the resource, heritage is the instrument and identity the goal.

There are three main sets of underlying assumptions, relevant to this argument that are only rarely questioned or even made explicit. First, there are those surrounding the belief in what can be termed 'localism', which is increasingly contrasted with its opposite 'globalism'. This article of faith holds that specific localities may or may not possess a quality called identity, which links the people who live, work, visit or even imagine them from a distance, to such places. Secondly, this quality confers a universal benefit in the form of a public good by its very existence, the more of it that exists the greater such a benefit and conversely its absence, absolutely or relatively, confers equally unavoidable and universally applicable disbenefits. Thirdly, it is assumed that this public good confers feelings of well-being in individuals, which may in turn be a factor in promoting physical and mental health.

This chapter is concerned with the first two propositions and specifically with the process whereby the quality of local identity can be identified, measured and mapped so that it can be created if it does not already exist and enhanced and promoted if it does. This process does not necessarily only employ what is defined below as heritage but heritage is in practice the most commonly used instrument. Such deliberate intervention is generally regarded as a legitimate and feasible task of, in the first instance, public sector agencies operating in the collective interest.

The Entities

Three words, each of which is subject to vagaries of multiple, imprecise and sometimes even contradictory definition are used in this argument to describe the three main entities or resources that are to be interwoven. Each of these poses difficulties enough and the difficulty will be compounded by the attempts to combine them.

Landscape has been traced to an idea evolved by painters from the sixteenth century for completing their canvases with a background depiction of a usually imaginary scene of nature (*landskip* in Old Dutch where the practice was first labelled). As a verb it acquired the meaning of aesthetically improving land (the Dutch noun *landschap* being transformed into the English verb landscaping). It has since been loosely extended to any combination of geomorphological and biological physical features, whether or not modified by human action and extended even to any environment or setting ('the social or economic landscape') or just a geometric figure that is wider than it is high. Here the argument will revert to close to the original meaning in which the accent is upon the representation of structures from a world external to the observer. It is therefore by definition a product of the individual imagination. Townscape is a derivation from landscape generally composed of more built than natural environmental features but the distinction between the two is rarely absolute.

Identity has two contradictory meanings. It is that which makes something or somebody uniquely different. It is also that which makes somebody or something the same. Both meanings, identity and identical are often used interchangeably. Identity and place identity are not the same although often elided (Ashworth and Graham 2005). People may identify with specific locations as with specific social and cultural groups and places may be used to articulate or manifest such group identities. Equally however they may not. Much social and even political identification has no need of place. It may even be that in an increasingly mobile world, local place identification, far from being a universal basic human need, is a preoccupation of an unusually place-bound minority. Geographers and planners in particular are at permanent risk of succumbing to a place fetishism that fails to recognise that much, if not most, human identity is not specially place-bound. It may even be necessary to note that the term 'sense of place' is often used but the sensing comes from the people not the places, as places have no sense. Quite evidently, identity is ascribed to not given by landscapes and townscapes. Stones, bricks, trees, rivers and hills are physical entities that have no intrinsic cultural values whatsoever until someone endows them with these. They are not passively waiting to be discovered, recognised and appreciated. The active variable therefore is the 'someone' and the unspoken justification, 'for some reason'.

Heritage is not an artefact or site associated with past times, conditions, events or personalities. It is a process that uses sites, objects, and human traits and patterns of behaviour as vehicles for the transmission of ideas in order to satisfy various contemporary needs (Ashworth 1991). It is a medium of communication, a conduit of ideas and values and a knowledge that includes the material, the intangible and the virtual (Graham 2002). Heritage is a product of the present that draws upon an assumed imaginary past in the interests of the present and of an equally assumed imaginary future.

Like all products of the human imagination heritage and identity will change through time as new presents supersede the old who then imagine and identify with new pasts and new futures. Heritage and identity are thus driven by current needs, fashions, and tastes, and no universal, eternal and inalienable heritage values exist (Graham et al. 2000). Our current sacralised artefacts, monument lists, inscriptions, collections and icons are just the fashions of the recent past fossilised into a different present. They are the attempts, fortunately largely futile, to colonise an imagined future with our contemporary values as we physically clutter it with our preserved artefacts.

Heritage and identity are therefore created as they are needed and landscapes re-represented accordingly. All three are demand driven and thereby at least theoretically ubiquitous and infinite. In theory you cannot deplete, over-use, or even run out of heritage or identity. An excess of demand over supply is just a result of specific time or place bound management failure to increase supply, which being a product of the human imagination is, at least in theory, infinite.

With all three of these entities the important element is not the physical or spatial components that constitute the landscape, the identity and the heritage but the deliberate and conscious process through which they are created. The three processes are those of representation, identification, which is an active process performed by an identifier and 'heritagisation'. These processes create landscapes, identities and heritages respectively. In all three the active verb is more important than the resulting noun. Thus in combination we arrive at 'heritage landscapes' as an instrument rather than a condition: 'heritage identity' whereby an identification is made between some aspect of a present and of a past: 'landscapes of identity' whereby the representation confers either uniqueness or sameness upon some observer. To reverse the propositions, landscapes neither reflect nor confer identity through any intrinsic attributes: heritage is not a transmission of any aspect of a past to a present or future: and identity is not an irrefutable endowment from a genetic or social donor. All three are contemporary creations for contemporary purposes. The important questions are thus; Why do we create them, who are 'we' in this context, what do we expect from them and do we achieve what we expect? These are planning questions relating to the needs of contemporary societies not questions relating to history or to inherent characteristics of the natural and built environments.

The Dividend

The objective of combining the three entities is the production of what is called here the identity dividend. This in turn is founded upon three main assumptions about space (a geometric form) and place (a cultural construction). The first is that space is occupied by a palimpsest of localities, which are demarcated by some common collective identity, a concept similar to that of collective memory and with similar difficulties namely, is this collective an aggregate summation of a myriad of individual identities or something quite separate and plausibly different? Hierarchy is an intrinsic characteristic of place, therefore is the 'Russian doll model' of comfortably nesting hierarchically related identities (Ashworth and Howard 2000), ranging in size from the single individual to the largest collective unitapplicable or are models of conflict or coexistence between levels in the hierarchy more relevant?

Secondly, there is also an assumption that space can be regionalised in a geography of identity. Places can be classified as identity-rich, identity-poor or even identity-less. Expressed so boldly, this is at least a highly questionable notion yet it is quite implicit in many public policies for the shaping of place identities.

Thirdly, there is also a central assumption that is rarely if ever explicitly explored, largely because the answers are assumed to be either self-evident or devoid of any useful meaning. This is that people need to identify with places and specifically localities. There is an assumption that a distinctive, clear place identity is, if not an absolute necessity for individual stability or fulfilment, at least beneficial and satisfying. Specifically, a dividend, whether economic or psychic, is expected to emanate from the officially endorsed 'identity-rich' regions and be automatically conferred upon their inhabitants and other users of such places. Conversely the identity-poor or identity-less areas possess less of this quality of place and may even be 'placeless'. This is a concept introduced by Relph (1976) in which the uniquely local is sacrificed for the universally familiar and thus the identity dividend is traded for functional benefits, as in a generically replicated shopping centre or airport departure lounge. Such places are thus assumed to have less, or even no such, dividend to confer.

This prompts three lines of thought. First, it may be argued that these assumed benefits accrue to collectivities and contribute to collective attributes such as social cohesion or political allegiance rather than to the individual who may receive little or no automatic benefit. Secondly, a strong place identity cannot be assumed always to possess desirable connotations: the identification can equally be with negative qualities. Dachau, Chernobyl and currently Basra have strong and instant identification as distinctive places. There is also the 'dirty old town' defensive, affectionate, or even 'pride in adversity' place identification of the 'we are the worst/poorest/roughest' variety. Thirdly, there is the equity argument. If Bath has more identity conferring more benefits than Slough (or Heidelberg than Gelsenkirken), then policy, not least for heritage, should be redirected towards compensating the deprived rather than rewarding the already fortunate.

There is something of a paradox in this reasoning once it is applied to daily life. On the one hand it is relatively easy to reveal and demarcate the existence of such a place identity value surface and relate it to the social and psychological conditions of inhabitants. However there is little evidence that people who live in identity-rich areas are happier or less dysfunctional than those who do not. Even if the inhabitants of Slough are discovered to be more prone to various psychological or social malaises than those of Bath, this discrepancy could more easily be explained by other economic or demographical variables.' On the other hand, individuals do seek out places that possess a sense of place as long as that sense is composed of elements of visual comprehensibility, aesthetic gratification, and agreeable historical associations. As tourist, people freely seek out and pay to visit such places: as resident they are prepared to pay a premium to inhabit places of such 'character', to use the real estate broker's portmanteau term for the qualities described. Thus while it is easy to question the causal chain that leads from heritagisation, through place identity to individual well-being, people frequently seem to behave as if such an identity dividend exists

The Planning

Many would regard the question, 'Can it be done?' as absurd because planners are doing it and have been doing it for some generations, at first implicitly and more recently quite explicitly through government policies at scales from the local to the global. However, it may be that what they are doing is not necessarily what they think they are doing, nor are the results consistent with the objectives. Planning, as considered here, is the deliberate goal directed intervention, usually by public agencies operating in pursuit of a collective public interest. Treating the planning of heritage landscape identity as a sub-set of this raises the question 'does a particular form of planning exist, or could be developed, that could undertake the task of planning for heritage landscapes, as defined above, for the benefit of the wider society?' Can planners shape the conditions, whether economic, social or morphological, that favour such place identities or, more modestly, avoid shaping those conditions that disfavour them? The phrase 'heritage landscape planning' appears in the literature but usually as a means of managing land use activities, or more usually a cynic might observe, managing their subsidy, from government departments. It may be that 'landscape planning' or perhaps more properly 'using landscapes in planning' is just an application of existing techniques and instruments in new approaches to intractable problems. Equally, even if it can be done, it may be no more than a particularly unique strategy applicable in a few chance places rather than a more general development strategy applicable everywhere and anywhere.

Such planning when self-consciously drawing upon imagined pasts, seems to operate within three different paradigms (Ashworth 1997). Preservation, is easily defined as the protection from harm whether natural or man-made for whatever reason. The planning implications of this are limited and essentially negative. Spaces and buildings are removed from 'normal' planning consideration. Preservation is therefore counterpoised to development and becomes the negation of planning. The shift to a conservation paradigm, defined succinctly by Burke (1976) as 'preserving purposefully' added two new dimensions namely, ensemble and purpose. Conservation did not replace preservation but was incorporated somewhat uncomfortably alongside it, with its notable achievements being area conservation and thereby the necessary addition of function to form and, consequently, the acceptance that development, meaning usually in this context the management of change, was incorporated into the process rather than excluded from it. The designation of areas and the decisive role of contemporary purpose allowed the planners and urban mangers to supersede the architects or historians who are reduced to mere resource providers. It was a logical progression to arrive at the heritage paradigm, defined as 'the contemporary uses of the past' (Ashworth 1991). This was however a radical shift whose implications are yet to be fully realised in many quarters, for two reasons. First it focussed on the present, with both the past and, as in all planning, the future being no more than imagined entities. Preservation assumes a real past can be preserved for a real future, whether a future wants it or not. It validates its selections on the slippery criterion of authenticity. Heritage knows only the present as real. Pasts do not exist and therefore cannot be preserved: they can however be imagined and this imaginary entity can be represented not least in townscapes. Secondly, use becomes the variable determining both inputs (selection, packaging and interpretation) and outputs (uses and users). Identity becomes an end-state, assumed to be a desirable contemporary attribute of places, to be achieved through planning.

The Belvedere Case

A single national case illustrates the operation of these conceptual paradoxes in planning practice. Four government ministries in the Netherlands initiated a joint policy in 1999, known in short as 'Belvedere'. The quite explicit objective was to be the discovery, protection and enhancement of landscapes, whether rural or urban, that were deemed to be significant in projecting and strengthening local identities. The simple and clear logic was, that heritage endows landscapes with meaning that reflects a specifically local identity. If such landscapes can be recognised, preserved from harmful developments and promoted through projects based upon these unique attributes, then this will strengthen local identity (OCW, LNV, VROM, VW, 1999a) and confer a dividend upon inhabitants and visitors alike. The most interesting and probably enduring action has been the inventorisation and designation of around 70 so-called 'Belvedere regions' and 105 cities with cultural-historic significance (OCW, LNV, VROM, VW, 1999b; Ashworth 2007). These now encompass around a third of the total land area of the country and includes every major city and almost all the minor ones as culturally historically significant. It is not clear what the implications are for places listed as having neither culture nor identity (eg. Lelystad, Drachten, Zoetermeer and Almere), which no doubt have a distinctive identity at least for their inhabitants. The planning method was encapsulated in the seemingly paradoxical slogan, 'preservation through development' (*behoud door ontwikkeling*), which called for active intervention rather than just passive protection through a spectrum of actions that range from more traditional monument restoration, reconstructive building, site marking and even regional branding campaigns. The intention is for central government to provide stimulation, policy guidance, seed-corn funding, and possibly expertise in support of local projects devised and executed by local public and non-statutory agencies and private firms and even individuals. To date some 200 such projects, all focussing on the enhancing of local identities through historicity for the benefit of diverse use groups, have been initiated and more are expected in the period up to 2010.

Belvedere in Theory

This policy and the programme it has initiated raises important and timely issues that have rarely been debated at this level of government seriousness before. It can be argued (Ashworth and Kuipers 2001) that the discussion is quite seriously conceptually flawed and ill focussed. It is not just that the goals are ill defined and probably impossible to evaluate but more fundamentally it misunderstands heritage, identity and in particular the existence of a dividend for society or the individual.

The policy assumes that, 'cultural heritage is a reflection of settlement history to inhabitants and tourists' (OCW, LNV, VROM, VW, 1999-a, p.11) whereas, as has been argued above, precisely the reverse is the case. Heritage is a reflection of what we have decided we want to reflect through the creation of imaginary pasts and their projection to equally imaginary futures. Place identity is dynamic not static so the Belvedere report cannot fulfil its intention of recognising, locating, and listing regional identities because these are not waiting to be discovered and then preserved for eternity. Secondly using heritage as part of a policy in support of local identities is to enlist a treacherous ally. 'Heritagisation', the process through which selected aspects of the past are re-presented for contemporary consumption, is itself a part of the very globalisation it is being recruited to oppose. Heritage places are created by global investment and development corporations, global designers and local politicians and planners inspired by global ideas and exemplars: all are reacting to global demands (Ashworth and Tunbridge 2003).

Similarly the idea that history endows regions, with a distinctive identity is treated as self-evident whereas identity is ascribed to landscapes not given by them. The active variable therefore is the 'someone' and different people at different times identify in different ways with the same places. It also assumed that local distinctiveness is declining, at least in the western world, as a result of an increasingly global economy and society and that this is self-evidently undesirable and should be counteracted by large scale government intervention. Belvedere claims to be offering some respite from the homogenised 'global village' in conserved oases of urban and rural localities with distinctive local identities. Even if economic globalisation must be accepted, cultural localisation can be, it is claimed, maintained. Thus localism is seen as a valued counterpoise to a regrettable homogenisation. Finally, not only are heritage and identity assumed to be singular homogeneous entities, so also is contemporary Dutch society. A place is assumed to have one single universally accepted identity that reflects a consensual set of cultural values, derived from history. It is doubtful if this was ever the case in the Netherlands, or indeed elsewhere in Western Europe, and, if it was, it has certainly long ceased to be so. Of course place identities have long been used by governing elites to convince the governed of the existence of such a consensus and thus the legitimacy of government (Graham et al. 2000). Replacing or supplementing nationalism with an assumed consensual localism or regionalism is just as politically exclusive and socially disinheriting. If societies are plural then so also is the identification of people with places and if heritage, and culture more generally, is the active ingredient in shaping diversity in landscapes and cityscapes then it is particularly undesirable that this is assumed to be itself monolithic.

Belvedere in Practice

Despite the conceptual misgivings, the national policy has since 1999 both evolved in the diversity and creativity of its scope and devolved into numerous and very varied local coalitions of official and unofficial agencies pursuing local interests. It is these local 'projects', rather than the original political intentions or subsequent definitional misgivings that characterise the Belvedere contribution to Dutch urban and rural planning and therefore are likely to deliver the identity dividend. Three very extensive exemplar projects were included early in the national policy to indicate the nature of government thinking and stimulate responses from others. These were, first the '*Limes*', the Roman boundary zone that stretched West-East across the entire centre of the Netherlands (Veenboer and Groote 2006), secondly the Waddenzee, the tidally inundated area between the Frisian islands and the mainland (the *Lancewadplan)* and thirdly, the mid-nineteenth century fortification system (the Nieuwe Hollandse Waterline).

Belvedere projects however are not only national scale visions covering major developments in wide parts of the country, they may also typically be very small scale and locally specific thus reinforcing a local identity. One of many such illustrates the principle of integration and cooperation between owners, state agencies, local governments, community pressure groups, commercial interests and individuals, in visioning, financing, executing and managing sites, for equally diverse user groups. Fort Asperen in the province of Gelderland was part of the nineteenth century defence works that included artillery emplacements, military buildings, water barriers and inundation equipment and zones that comprised the *Nieuwe Hollandse Waterlinie*. In essence the project comprises a single disused and decaying building and its largely inaccessible surroundings, which are to be both restored and developed to serve a number of new functions, including heritage interpretation, an arts centre and outdoor recreation facilities. The development partners involved include the land owners (the Forestry Commission), a foundation established to manage the complex, the provincial and district local governments, the national governments agencies for monuments, the arts, culture and education and a number of local interest groups. The financing is equally diverse with the initial 'vision' being funded by Belvedere itself with the modest contribution of EURO 40,000. Building and landscape restoration, facility development, maintenance and management are budgeted at more than ten times this sum and allocated from already existing local and national government funds. To a large extent Belvedere therefore is a reallocation, prioritisation and focussing of existing resources with some commercial leverage.

Once stripped of the political rhetoric of localism and anti-globalisation, what is left is a mix of unconventional historic preservation and restoration, landscape enhancement and protection, outdoor recreational management and heritage creation and marketing. The most notable feature in comparison with such policies for the last 30 years is the long term nature of the planning, the wider spatial scope and coverage and the cooperation of a wide diversity of the parties involved in financing, planning and operation.

From Heritage to Health

The attempt here to demonstrate the existence of a causal chain from heritage creation through place identities, to a dividend of well-being that contributes to health is clearly incomplete and at points ambiguous. Heritage is a major instrument through which places are created for the satisfaction of contemporary needs, amongst which is local place identity. Heritage is seen by policy makers as unique and its use will therefore automatically reinforce the uniqueness of the place. This place is different and special, and by derivation we who inhabit it are also different and special, because our history expressed through our heritage is necessarily different and special. It is this idea that lies at the heart of many if not most local policy initiatives for shaping heritage places. It is also possible to demonstrate that people are prepared to pay a premium to experience some types of place identities, which in turn suggests that some form of collective well-being is bestowed by such places. However it is also clear, even from the arguments raised above, that this process is more complex, more multifaceted and especially less predictable, than is generally assumed by those attempting to do it.

The attempt of national governments to support a sense of local identity may lead to a standardisation of what is conceived and planned to be local that is itself homogeneous. The practice of public heritage creation can be as much global as local. The local may become global in its reproduction of the same local features and conversely the global may itself be a universalisation of what was originally local (Ashworth and Tunbridge 1990; 2000). The concept of collective heritage, like the notions of collective memory or collective identity, with which it strongly related, needs more careful consideration than it generally receives. The collective is an imposition from outside for some purpose external to the individual, always prompting the questions, 'who is doing this and for what purpose?' The peoples, the identities, the images and the purposes are all usually too plural to be reduced simplistically in this way. Similarly if the consumers are plural so also are the producers. Heritage is created and transmitted by a variety of official, state-sponsored agencies in pursuit of public policies and also by unofficial forms, commercial and private agencies for non-public purposes: the latter often being different from, and even subversive of, the former. As with many areas of policy, including particularly those relating to the conservation of the natural and built environments, the question of who is making decisions becomes intertwined with the decisions themselves. In this case, 'who is the identifier?' may be as important to determine as 'what is identified?' This returns to the issue discussed in various guises of insiders and outsiders. There is certainly the possibility of the absurd situation where outsiders define the sense of place of locals who are informed what their recognisably distinct local identity is to be, which would seem to defeat the initial purpose of an exercise in localism. More subtly, there is the distinct possibility of an interaction between the two place identities as projected for external consumption and those intended for local internal use. Outsiders may seek out aspects of the local identity for various reasons while locals similarly adopt the externally projected images of themselves. Place-product commodification and branding cannot be separated for a perfectly segmented market. In reality local insiders and non-local outsiders, official policies and local reactions, diverge and converge in a continuous *perpetuum mobile*.

The search for a single collective place identity is a chimera and the undertaking of such an enterprise may deny the social and ethnic diversity of contemporary society. Similarly the local benefit assumed to accrue as a result of such policies, the 'identity dividend' is neither inevitable nor if it exists, likely to be equitably bestowed. Communities and their identities are quite obviously in a process of constant change and are not static entities capable of being frozen at a particular moment in time and space. The physical elements of landscape can be physically and legally protected but the local community that inhabit them and the local identities that they represent cannot. Thus heritage will itself be in a constant state of reselection and re-interpretation by the changing societies that are creating it in response to their changing needs. If there is one single lesson of this argument, it is that words such as identity and heritage, when used in public policy, must almost always be pluralised.

References

Ashworth, G.J. (1991), *Heritage Planning: Conservation as the Management of Urban Change* (Groningen: Geopers).

—— (1997), 'Conservation as preservation or as heritage: Two paradigms and two answers', *Built Environment* 23(2): 92–102.

—— (2007), 'Heritage places and place identities', in G.J. Ashworth, P. Pellenbarg and P. Groote (eds), *A Compact Geography of the Northern Netherlands* (Assen: Boekvorm Uitgevers).

Ashworth, G.J. and Kuipers, M.J. (2001), 'Conservation and identity: a new vision of pasts and futures in The Netherlands', *European Spatial Research and Policy* 8(2): 55–65.

Ashworth, G.J. and Graham, B.J. (2005), *Senses of Place: Sense of Time* (Aldershot: Ashgate).

Ashworth, G.J. and Tunbridge, J.E. (1990), *The Tourist-historic City* (London: Belhaven).

—— (2000), *The Tourist-historic City: Retrospect and Prospect on Managing the Heritage City* (London: Elsevier).

—— (2003), 'Whose Tourist-Historic City? Localising the global and globalising the local', in M. Hall (ed.), *Globalisation and Contestation* (London: Routledge).

Burke, G. (1976), *Townscapes* (Harmondsworth: Penguin).

Graham, B.J. (2002), 'Heritage as knowledge: capital or culture', *Urban Studies* 39: 1003–17.

Graham, B. Ashworth, G.J. Tunbridge, J.E. (2000), *A Geography of Heritage: Power, Culture and Economy* (London: Arnold).

OCW, LNV, VROM, VW (1999a), Belvedere: beleidsnota over de relatie cultuurhistorie en ruimtelijke inrichting. Den Haag: VNG-uitgeverij.

OCW, LNV, VROM, VW (1999b), Belvedere: beleidsnota over de over de relatie cultuurhistorie en ruimtelijke inrichting. Bijlage: gebieden. Den Haag: VNG-uitgeverij.

Relph, E. (1976), *Place and Placelessness* (London: Pion).

Veeneboer, G. and Groote, P. (2006), 'Kan Belvedere de Limes weer tot leven wekken', *Geografie* September: 37–39.

Conclusion

Allison Williams and John Eyles

In this final chapter, we do not want to reflect so much on the chapters in the collection but point to some trends in the ways in which senses of place and health may be conceived and shaped. Some of these are of course embedded in the substantive contributions in the book. DeMiglio and Williams, for example, point to the challenges of ever defining the complex and nuanced relationship between senses of places and health with the shifting lenses of the academy and interests of policy-makers and practitioners. Relph looks at philosophies of place and experiences of multiple places in the context of twenty-first century environmental challenges in which considerations of health are more implicit than explicit. Stefanovic argues that a holistic approach to studying sense of place enables a broad view of health for its incorporation into environmental decision-making. Manzo highlights the importance of psychological processes and characteristics when she emphasizes the importance of housing and home for place attachment and well-being. Santana and Nogueira see sense of place as part of the social characteristics of individuals and communities for the reduction of unhealthy behaviours in neighbourhoods. Williams and Patterson show the relevance of incorporating considerations of nature and leisure into place and well-being studies – in some ways they point again to a possibly inseparable linkage between place and time in understanding the human condition and well-being. Carreras, through portraying the senses of place of youth through their mental maps, writes of their detachment from their local societies and their perceptions of healthfulness in the emigrant's world. Krevs, in looking at sense of place themes in post-socialist societies, provides examples of its relationship to quality of life/well-being through emotional states such as love, hate and fear. Buzzelli introduces environmental justice as a key concept and uses it to reveal senses of place in neighbourhoods defined by poor health and environmental degradation. Ashworth extends the idea of sense of place to include heritage identity, arguing that this is potentially an important dimension of place identity and quality of life. Through these cases the importance of sense of place for well-being emerges but it is variable and context-dependent. Some of the contributions also highlight the fact that this relationship can change over time as well as between different groups. If, therefore, sense of place and health/well-being are changing, we need to ask how are they changing and how might those changes affect their investigation and treatment in the policy arena? We answer this question by first exploring the changing nature of place, then the changing nature of health, after which we provide a number of suggested directions for further investigation. Furthermore, we suggest that this examination begin with adverse health outcomes, although we recognize the importance of examining therapeutic landscapes for researching positive health outcomes, the latter which are briefly touched on below.

Although we have focused mainly on negative health outcomes in this final chapter, we know that some environments can be positive for health such as what Jutras (2003) found for the well-being of children, and Gross and Lane (2007) for private green space and gardens in particular. Certainly a number of contributing chapters in this volume suggest the connection between sense of place and positive health outcomes. DeMiglio and Williams review the evidence supporting this relationship. Further, Manzo illustrates how public housing residents laid down roots, formed place attachments and created bonds of mutual support with neighbors – all of which impacted positively on resident's well-being and were missed once displacement occurred. Williams and Patterson, in their examination of recreation in natural settings and the place meanings and attachments associated with natural landscapes, suggest that leisure and time use are important dimensions of health and well-being, Sense of place has long been identified as central to the therapeutic landscape concept, which provides a framework for analysis of natural and built, social, and symbolic environments as they contribute to healing and well-being in places – broadly termed landscapes (Gesler 2003). Although there have been a number of critiques which suggest, for example, that places can be simultaneously healing and hurtful and not limited to places celebrated for their reputed healing qualities, the wide-ranging and ever-expanding use of the concept both within and outside geography suggests that the positive health outcomes with which they are associated have been resonating widely.

Changing Nature of Place

Although there are many of us who romanticize the times of old when people lived for generations in the same farmstead or village, many of us have little, if any, historical memory of such an existence, if even based on oral family histories. As citizens of the 21st century, many of us have lived in more than one place, whether the result of war, social discrimination, economic or work-related reasons, family reasons, the pursuit of quality of life, or a combination of the above. This enhanced mobility that has increased over the generations has most certainly affected sense of place perceptions, allowing us to experience multiple places, albeit in a much more less intense manner when compared to that experienced by our forefathers less than an century ago. Further, technology continues to have a growing impact on the way we live our lives in various places. Technology not only allows us to experience a number of different places simultaneously, such as those in real time and those available virtually, but has affected the speed in which we are expected to live our lives, potentially leaving little room to experience what Relph (2007, p. 17–18) has called the spirit of place, or the 'distinctive identity of somewhere', whether that be defined by 'natural landmarks or remarkable built forms'. In other words, our quick-paced lives – influenced by a growing number of technological tools and mediums, may leave little time, energy and interest 'to grasp and appreciate the distinctive qualities of places'. Some would argue that enhanced mobility and technological innovation are both contributors to and outcomes of globalization. Some believe that globalization, or the growing interconnectedness experienced by both people and places as a result of the global convergence of a number of systems – the economic, cultural, political, religious and social, has left the world both faceless and placeless. If this is true, then we as a society need to be more intentional about nurturing sense of place given our human need to be connected – in some quantity and quality, to place.

We know that place is and has always been a process and is, thereby, dynamic and changing. The last ten to fifteen years has seen a number of key drivers that are changing the nature of place more quickly than ever before. Although there are likely many more, the three central drivers introduced above and commonly discussed in both the academic literature and popular media are: mobility, technology and globalization. Each of these will be discussed in turn, beginning with globalization, given its broad reach.

Globalization

Globalization is best known as an economic phenomenon, playing a central role in the emergence of global markets of production and distribution and in world-wide financial markets. The realization of a global common market, where both capital and goods are exchanged worldwide, has been central to what many understand to be the hallmark of globalization, although it is argued that this financial element is not globalization per se, but rather liberalization, often occurring concurrently with globalization but not inherent therein (Scholte 2005). Indeed, processes of globalization extend to systems beyond the economic, and include changes to cultural, political, religious and social systems. There is no doubt that the growing interconnectedness of people and places are driving the changing nature of place.

Amin (2002: 385) advocates for a broad view of globalization, in which 'the spatiality of contemporary social organization' is the central tenet, and points to the association between the actions of large global forces and meanings of space and place. Similarly, Scholte (2005) describes globalization from a social lens, understanding it as an alteration in social space and broadening of human connectivity. Although many of the economic and political decisions are, in particular, made by far-reaching global institutions – such as the International Monetary Fund or the World Trade Organization for example, all of the effects of globalization – from the economic through to the cultural, social and ecological, happen in local places (Meyer et al. 2006). For example, the global environmental challenges of climate change, which are arguably impacted by increased trans-global trade, the resultant resource expenditure of these activities, and related trade policies (ICTSD 2006), are experienced by all local places. Whether evident in seasonal temperature changes, drought or natural disasters, these challenges require not only international cooperation but local action. Local action at the population and systems levels calls for the assessment of local vulnerabilities, resulting in both responses to address the environmental challenges, as well as improved security and control in place (Bentley 2007; Menne and Bertollini 2005, 1284). Another example is the local experience of globalism, which encompasses the experience of world culture through cultural diffusion – via products, ideas, technology and practices; to illustrate, it is not uncommon for a person to be working on a Sudoku puzzle and listening to loud African drumming on their Ipod while waiting for their Mexican-Indian fusion take-out eaten before attending their weekly Latin bop class.

There is a clear demarcation of ideas between those who are pro- and anti-globalization (Chomsky 2000), with some of the latter feeling that their sense of place may be threatened. This fear has the potential to breed "defensive and reactionary responses – certain forms of nationalism, sentimentalized recovering of sanitized 'heritages', and outright antagonism to newcomers and 'outsiders'" (Massey 1994, p. 1). Further, others have suggested that the negative effects of globalization are many, including social disintegration, increasing poverty and alienation (Capra 2002) – all of which have obvious impacts on sense of place. Although terrorism is not understood to be a direct impact of globalization, some have argued that one of the outcomes of globalization is growing income inequality within and between nation-states and more generally across the First World and Developing World contexts (Goldberg and Pavcnik 2006). While recognizing the complexity of this relationship, some propose that it is this income inequality that has been and continues to feed terrorism and revolt (MacCulloch 2005). As with many social movements, terrorism has been using enhanced mobility and technological innovation to its advantage, more quickly and effectively propagating the fear that it intentionally produces. It is likely that those on the receiving end have responded to this fear by digging their heels further into what they call their place. Such a reaction could be understood as an illustration of territoriality/protectionism or as a need to find solace in stability of place.

Massey (1994, p. 9) suggests that a progressive, outward-looking 'global sense of the local, a global sense of place' is one way to 'hold on to that notion of geographical difference, of uniqueness, even of rootedness if people want that, without being reactionary' (p. 5). Such an outlook celebrates cultural difference – in all its many forms, while participating in the social relations that are inherent in place. Further, this approach is 'extroverted and includes a consciousness of its links with the wider world, which integrates in a positive way the global and the local.' (Massey 1994, p. 7).

Globalization, in its broad characterization as an alteration and expansion of human spatiality, has the ability to have wide-reaching health impacts. As the effects of globalization in its various manifestations are dependent on an immeasurable number of interdependent factors, so too are the resulting health effects of the phenomenon. There are many direct ways in which globalization can influence health. Certainly increased global mobility (discussed below) and trade have facilitated the spread of infectious disease, a movement that is particularly devastating when disease agents are carried to populations within which there has been little exposure and thus low immunity (Feachem 2001; Kawachi and Wamala 2007). In contrast, however, the advancement of and increased access to new information and communication technologies has contributed to improvements in disease control and prevention, through enhanced knowledge translation and health surveillance activities (Lee and Whitman 2003; Kawachi and Wamala 2007).

There are also numerous indirect, and sometimes more obscure, ways in which globalization is entwined with health pathways. Labonté and Torgerson (2003) consider various frameworks in which relationships between globalization and health can be analyzed. Interconnected elements of globalization that impact health outcomes range from (but are not limited to) trade agreements, Official Development Assistance (ODA – donor country support) programs and related conditions, environmental protection policies, environmental resource bases, civil society engagement, subsistence production, education and health behaviours (Labonté and Torgerson 2003: 13). These span spatial scales from the global to the local. More recently, the ways in which globalization is related to social determinants of health (SDH) have been explored. Pathway clusters within which globalization can impact SDH include trade and financial liberalization, production reorganization, debt, urban growth and restructuring, resource exploitation and environmental, and marketization of health systems (Labonté and Schrecker 2007). A focus on distribution and health equity reminds us that these pathways can contribute to positive or negative health outcomes, the negative felt mostly forcefully by the already impoverished (Labonté and Schrecker 2007).

Technology

In many ways, technology has been globalization's toolbox. Technological innovation has and continues to happen at break-neck speeds, ranging from the advances made in travel modes through to the digitized world of the new economy which allows work to be distributed and completed anywhere (Scholte 2005). While being a key instrument in time-space compression, the central outcome of the availability and use of innovation technologies, among which include the Internet, communication satellites, wireless telephones and submarine fibre optic cables, has been the development of a worldwide telecommunications infrastructure. This global telecommunications infrastructure has been working in tandem with nation-states facilitating open, transborder data flow; this has resulted in instantaneous and simultaneous information exchange, producing an unprecedented availability of information which is manifested as what Scholte (2005) refers to as *supraterritoriality*. Such instantaneous information flows have become essential for worldwide communication more than ever before as boundaries between places recede, with information flows increasing between geographically remote locations (Feachem 2001). This is best illustrated in the flow of financial transactions in the wake of a global common market, which is characterized by the freedom to exchange either capital or goods practically anywhere.

The availability of technology and thereby its use is place dependent due to the fact that it is expensive and requires investment in certain resource infrastructures. Simply put, the have-not nation-states have minimal access to many of these innovative information technologies compared to well-to-do nation-states (Narayan 2007; Hafkin and Taggart 2001) which are the technological innovators and have pioneered the global telecommunications infrastructure; it is not surprising that these same nation-states have consistently faired well in this globalizing era.

So what impact have these technological innovations had on place and sense of place? Although somewhat dated, Kotkin and Siegel (2000) provide an extensive review of how, in the digital-age, the city and countryside are being 'remade'. Using a number of case examples across the United States, they suggest that it is the rise of the new (digital) economy that is making or breaking places – creating a new geography of economic life.

> ... this shift to a cyberspace – the increasingly global nature of business and ever more sophisticated telecommunications technologies – has made it easier for companies and elites to ignore their responsibilities to a particular city or place ... (p. 3)

The digital economy is made up of a high-tech workforce, the majority of which have nonmanufacturing jobs that can be carried out anywhere given that technology allows work to be distributed anywhere. Kotkin and Siegel (1999) suggest that places that want to succeed in attracting entrepreneurial and technical talent must devote their energies to quality of life concerns, which have risen in importance over the traditional 'economic-development drivers such as abundant land, low taxes, and housing and labor costs' (p. 4). Attractive places have become the 'new locational paradigm' in the geography of industrial location, leaving many places to fall behind in population numbers, economic development, and quality of life indicators, to name a few. Multinational corporations' interest in minimizing labour costs has seen many manufacturing and service activities move from the First World to nation-states within the Developing World context. One example of such outsourcing is Microsoft's technical support – a skilled knowledge service traditionally made available in the US, now located in Bangalore, India.

The impact of a virtual economy and what appears to be a growing virtual world for many who live in the First World context is what Relph (2007, p. 24) has named 'A sense of virtual place', which will 'involve many senses and emotions because it is mediated electronically ... will vary between individuals and ... have a community expression.' Rather than replace a real sense of place, he suggests that such a 'virtual place can be considered a variant of and an addition to the current distributed sense of real place that simultaneously acknowledges geographical diversity and seeks ways to make places with compelling identities.' In line with this extension of place, Green et al. observe the alteration of *spatial conditions of existence* as characteristic of new information technologies (2005: 805), although it is important to note that regardless of the sophistication of and ease of access to new space creating technologies, 'the localness of experience is a constant, and the significance of locality persists' (Meyrowitz 2005).

Mobility

The growing mobility of labour across the globe has been a very uneven driving force behind the changing nature of place, and particularly urban places. Its incidence is so great that it has been identified as one way in which the economic flows that characterize globalization can be measured; net migration rates and inward or outward migration flows (weighted by population), for example, have been used to measure globalization (Asongu 2007). Massey (1994) suggests that this enhanced mobility is being differientially experienced by those in control of it (voluntary) and those being controlled by it (involuntary). Those in control are those whose travel is voluntary, often chosen for career advancement. Those not in control are being moved around by a wide range of forces, including economic, political and social. These include migrant workers, refugees, and various types of domestic help, for example.

Mobility across the globe has never been greater with increasingly accessible travel and the freedom for citizens of many nation-states to travel internationally. Although this enhanced global mobility has brought about problems, such as the rapid spread of severe acute respiratory syndrome (SARS), it has greatly enhanced the tourism industry, immigration flows, and international cultural exchange. Movement within nation-states is increasingly common as well. Those living in the First World context intentionally move house for various reasons, as discussed. The average American, for example, moves house every five years (Dauncey 2003). Within the Developing Nation context, the pursuit of work and wealth is driving urbanization and resulting in the growth of urban slum environments.

So what effect has enhanced mobility had on sense of place? As suggested above, it has likely diluted sense of place perceptions for any one place, given that few of us have much place longevity given that we have not lived or come to intimately know any one place for an extended period of time. At the same time, experiencing multiple senses of place may instil a particularly strong sense of place-identity with one particular place, given that any one individual has one or more places to which they can compare; for many this one place is what they call home.

Reasserting the Local in Sense of Place

Although these three central drivers to place change are very much at work, we as human beings commonly have a single local place with which we live out the majority of our lives and with which we most identify. As Relph (2007) reminds us, place always has and continues to be central to the human experience:

> A life without places is as unimaginable as a life without other people. We all were born, live and will die in towns, neighbourhoods, villages or cities that have names and which are filled with memories, associations and meanings. Places are so completely taken for granted that they need no definition. They are the complex, obvious contexts of daily life, filled with buildings, cars, relatives, plants, smells, sounds, friends, strangers, obligations and possibilities. (Relph 2007, p. 17–18)

For most of us, this single local place, although changing like all other places, is full of placefulness. As such, our sense of place in this local environment has been shown to be useful in a wide range of contexts, from ecosystem management (Cantrill 1998; Williams and Stewart 1998) through to education (Sobel 1998) and administration (Driscoll and Kerchner 1999). The local provides not only an environment where we as individuals can assert our interest and investment in place, but also reap the benefits of a strong sense of place. Although there are innumerable examples that illustrate this relationship to local place, one unique case example is the ecomuseum concept, a concept which developed in France late in the 1960s and now is a worldwide phenomenon (Davis 1999). Unlike the traditional museum, ecomuseums are ideally dynamic, evolving developments, resulting from the ongoing interplay between the community and its environment. The tasks specific to education, preservation and collection are performed via community involvement and within the local context of the natural and built environment. Rather than being located in one place, ecomuseums are 'museums without walls', with a number of sites dispersed throughout the local place. Specific sites of importance can be defined by, for example, collective memory, natural history, human heritage, contemporary life, natural geological landmarks and remarkable built forms. An ecomuseum is an evolving entity that both feeds into and produces local sense of place and place identity through local people playing a role in its evolution. Now we turn to a discussion of how the nature of health is changing. We do this by having broken apart the WHO definition in this final chapter, and in so doing, emphasize promising lines of inquiry for the physical, mental and social dimensions of health.

Changing Nature of Health

For many years, the WHO (1946) have articulated a view of health that has gained broad acceptance. Health is defined as "a state of complete physical, mental and social well-being and not merely the absence of disease or infirmity." International work on "Health for All' together with academic developments, have resulted in this view becoming more goal- and action-oriented. Thus the Public Health of Canada (2007) sees health as a capacity or resource for living and the WHO Agency has based much of its work specific to the social determinants of health and population health framework on Frankish et al. (1996) who suggest that health is the ability that people have to manage the changes and challenges that life brings.

This approach has come under some criticism for being too overarching and sweeping, and for being very much a product of its time. Doll (1992) argues, for example, that including social and psychological components was conceptually important but has created practical problems. Saracci (1997) suggests that the origins of the definition just after the Second World War point to the linkages between health and peace. Furthermore, the state of complete well-being idea has more in common with happiness than health but has the effect of eliding health with happiness. This elision, Saracci argues, has consequences which may make the understanding of health issues more problematic than they need be. First, any disturbance to happiness may be seen as a health problem. We may argue that this has the potential to extend the medicalization of everyday life and see medical solutions to social and possibly community problems. Szasz (2007) has recently brought together papers that point to the medicalization of mental health, drug use, and end of life. While there may be disagreement on what should and should not be medicalized, Smith (2002) notes non-disease is as difficult to define as disease, with ageing, loneliness and ignorance often being seen as medical conditions when perhaps they are not. Secondly, as the quest for happiness is boundless so can be the search for health, seen perhaps best in the denial of ageing and mortality itself. Gillick (2006) looks at some of the consequences of this denial in terms of diagnoses, expenditures and treatments which may, she suggests, lead to poorer rather than improved elder care. Thirdly,

Saracci points to the fact that annexing happiness to health and regarding health as a universal human right suggests a particular form of happiness should be established with scant regard to personal autonomy. Finally, he argues that trying to attain happiness for all through health is not attainable but may lead to the use of resources on 'health' matters, detracting it not only from other arenas such as education and environment, but negating our abilities to differentiate between health needs and to achieve greater access, equity and justice in health for all citizens. Indeed we may note how in developed countries, health care expenditures continue to grow at what seems an unsustainable rate and how in many developing countries expenditures on tertiary care has tended to limit their abilities to invest in health-enhancing, disease-reducing strategies for all their populations.

If we accept Saracci's arguments we seem therefore locked into a view of health that is conceptually rich but perhaps of limited empirical use. Is this the case and what can we learn from our chapters? The WHO definition and its amendment for action have been useful. It has enabled a far broader understanding of what people experience on a day-to-day basis and what individuals, communities, governments and other agencies can do to mitigate problems and improve conditions that impinge on well-being. Furthermore some of the chapters have introduced philosophical ideas that point to the importance of health and well-being as rights to be obtained. Further, social and environmental justice considerations suggest differential treatment or access may be required to ensure the right to good health for all.

So how should health be seen? Saracci himself argues that health is a condition, similar to Frankish's idea of a capacity. In other words, health is a means in this view. If it is, however, a means, we need to ask to what ends and what outcomes are interests are. Thus, to move the agenda of sense of place – health research and practice forward would require us to specify what our purpose is, what our research questions are, and what are our outcomes of interest? Thus our view of health may be broad conceptually but it may become quite narrow empirically as we focus on specific dimensions of everyday life and experience. Do we thus move forward on many fronts as our authors indeed try to do? What then seem to be promising fronts?

For these suggestions, we have broken apart the WHO definition in this final chapter. In other words, we pay specific attention to the outcome of interest and while recognizing that there may well be a coincidence of outcomes (physical, mental and social), feel that the choice or emphasis on one is important for conceptual and empirical clarity. Furthermore, although we recognize the importance of the positive dimensions of health for well-being, we emphasize ill-health and disease. While defining disease is not without difficulty, such an approach decouples the link between health and happiness.

Physical Health

What then are the contemporary debates about the nature of physical health in so far as they relate to our sense of place-health interests? The outcomes of interest remain of course largely the same. We continue to be concerned with the major causes of death and disability – cancer, heart disease, diabetes, respiratory problems, accidents and so on. What has, however, begun to change is the factors which are regarded as potentially important in determining the presence of these diseases as well as their differential distributions between groups and places. Much of this work is beyond the scope of our project as are many of the recently identified environmental insults to human development and healthful living, such as trihalomethanes in drinking water and endocrine-disrupting chemicals in specific locales and substances. Indeed for illness linkages to become important for sense of place, they must first be related to environment and place. Thus for example, Longnecker and Daniels (2001) have shown some relationship between environmental contamination and the etiology of diabetes, particularly nitrates and nitrites for type-1 and biphenyls for type-2. Furthermore, an association has been found between Canada's Great Lakes areas of concern and hospitalization for diabetes (see Elliott et al. 2001). Diabetes may also be associated with places linked to particular lifestyles as has been found within First Nation reserves in Canada (Martin and Yidegiligne 1998). But the links to place are tenuous as they remain with heart disease and cancer. For heart disease associations have been found with environmental contaminants in Sweden (see Rosenlund 2005, Sergeev and Carpenter 2005). But the pathway to heart disease is complex. In fact stress has often been suggested as an intervening variable between place characteristics and heart disease (Bertazzi et al. 1989; Mathur 1960). We will examine this mediating factor again in relation to psychological health. For cancer, Danaei et al. (2005) point to the importance of lifestyle factors in the causes of cancer worldwide yet broadly defined environmental forces as far more important than heredity, at least for breast cancer (see Couto and Hemminki 2007). Hemminki et al. (2006) note that environment seems more important than genetics in explaining cancer incidence, although they note the potential significance of gene-environment interactions. Place is also implicated in the existence of many elevated cancer rates; Clapp et al. (2006) conclude in their review that farming states and cities, for example, are linked to elevated cancer rates. In some cases, the increased place-based risk seems small as Roberts et al.'s (2006) study of waste incineration shows. But let us note where place and its importance to us may be implicated in cancer causation. Passive smoking or environmental tobacco smoke (ETS) is seen as significantly increasing the risk of lung cancer among non-smokers (see Hecht 2006, Raposo et al. 2007). This exposure occurs in workplaces, bars and homes, the last two being significant havens in the world. So the home in particular – and not just for ETS – may become a dangerous place for physical health despite its positive nature.

Mental Health

With psychological health, there seems a more explicit role for sense of place, if place itself is relevant. There are again many mental health/illness outcomes that can be chosen, many relying on diagnostic criteria or clinically established cut-off points. But, as with physical (ill) health, it may be best to establish the salience of sense of place with adverse outcomes. Our present task, however, is to point to the role that psychological factors might play in the sense of place-health relationship. The unanswered question is posed well by Kubzansky and Kawachi (2000) when they ask how exogenous social conditions impact physiological health? They point to the pathway of emotions and the physiological, cognitive, and behavioural responses they evoke. In their paper, they examine anger and its potential significance in coronary heart disease as well as anger and depression. They point out how the combination of many of these emotions – anger and anxiety in particular – can raise blood pressure and lead to heart problems. We should note that all these emotions can be connected to place, especially places under threat or known to have a toxic environment. Eyles et al. (1993) discovered high levels of anxiety after the Hagersville tire fire. Thapa and Hauff (2005) found depression and distress among displaced persons in Nepal. Few if any of these studies formally link these psychological states and changes in well-being or health to sense of place and the effects of the event on that.

Kubzansky and Kawachi (2000) prefer to deal with emotion rather than stress, the latter which they see as poorly defined and wrought with many empirical problems. Stress in this view may be regarded as a black box. But Brunner and Marmot (1999) regard stress as an important way of seeing the relationship between an individual and adverse events in his/her environment. They point to the biological pathways that can help explain the relationship between the stressor and adverse health outcomes, as mediated by psychological characteristics. This cutting edge field is beyond the scope of our argument, but Brunner and Marmot (1999) point to how proxies for stress – isolation, distress, fatigue – may help explain different heart disease rates in western and eastern Europe. Wilkinson (1996) also shows how these psychological factors are necessary to help understand and explain why those in poor housing suffer higher rates of cancer and heart disease than those in good quality housing. It is then perhaps in these characteristics of stress itself that we may find the relative importance of place and sense of place for health. The variable relation between different types of places and these characteristics is being established. We have already referred to place disruption and indeed this may in some ways be the limiting case for this relationship with forced removal at the extreme (see Billig et al. 2006). Furthermore the growing neighbourhood and health literature (see Chapters 1 and 2) points to perceived problems in local areas being associated with poor metal health states. In fact in a systematic review, Troung et al. (2006) found individual-level factors, including psychological characteristics, as being far more important than area-level ones in predicting health differences. Matheson et al. (2006) report that the stress of living in neighbourhoods associated with material deprivation and high levels of residential mobility is associated with higher levels of depression. Most studies emphasize adverse health outcomes, such as the recent work in Quebec City (Pampalon et al. 2007). In these studies, however, it is not just the relationship between place and psychological health that is examined. The nature of the local social environment itself is also investigated. Thus prior disrupted events and the lack of local institutions for social cohesiveness and a sense of belonging are part of the development of hopelessness and distress among youth living in inner cities (see Bolland et al. 2005). The social dimension of health/well-being is also present.

Social Health

As Berkman and Glass (2000) point out, it is now more than 20 years since the first studies examined the influence of social relationships on mortality. It is true to say that the impacts of social networks, social cohesion, social inclusion and social capital (positively and negatively conceived) on health has become a major academic industry. There are now several reviews of social capital (and its related concepts) and health. Luke and Harris (2007) examine it in relation to public health, Kunitz (2204) for health in general and De Silva et al. (2005) for mental health. Some reviews are critical. For example, De Silva point to the differences between individual and ecological social capital. Both Hawe and Shiell (2000) and Carlson and Chamberlain (2003), for health promotion and health disparities respectively, point to its conceptual difficulties. Indeed while there have been significant attempts to link social capital to place and health (see Lochner et al. 1999, Mohan et al. 2004), much recent research has tried to sort out some of the conceptual and empirical difficulties. For example, Bowling et al. (2006) fail to show measures of social contact being independently related to health at the local level while Ferlander (2007) concludes that it is important to distinguish between bonding and bridging social capital as they imply different resources, support and obligations. Much attention has been focused at the neighbourhood level on bonding capital (e.g. Brisson and Usher 2005, Poortinga 2006). Yet there remain contrary voices. Kushner and Sterk (2005) suggest social cohesion may just as easily lead to poor behaviours and health. Welshman (2006) argues that there is a normative dimension to social capital which he addresses by relating it historically to the underclass and the culture of poverty.

These voices do not deny the existence or importance of social capital or cohesion. They point to the need to understand their constructs. The importance of trust and control over one's life and environment may be the key. Araya et al. (2006) certainly find this for mental health outcomes and social capital. Sampson et al. (1999) point to the role of shared expectations about informal local social control and reciprocal exchange, not only for well-being but for the willingness to intervene in neighbourhood affairs. Sometimes local ties and social cohesion may act against this willingness to intervene as Wilkinson (2007) found in two distressed New York City neighbourhoods. Such adverse conditions may lead to stressful living, perceived neighbourhood disorder and depression (see Ross, 2000) or anomie (see Fullilove et al. 1998). These emotions would appear to work against the collective efficacy that is seen as so important in reshaping neighbourhoods and improving health outcomes (see Sampson et al. 1997). Indeed this is an important intervention point for sense of place studies. How is the place seen? How do changes or daily living itself affect these senses? How do they vary between groups in the same neighbourhood and between neighbourhoods? Studies of place indicate that if disorder is perceived in a local area, resident self-esteem decreases, their stress levels increase and their willingness – as well as that of the municipality, to intervene decreases (see Ross et al. 2001, Haney 2007). As Sampson and Raudenbush (2004) comment, seeing disorder may lead to neighbourhood stigma and adverse health outcomes. It is also imbued with social meanings that may generate self-reinforcing forces that perpetuate the conditions themselves. And these conditions are likely to be disadvantageous to health with respect to a variety of conditions (e.g. Cohen et al. 2000, 2003) and result in greater inequities in health status (see Frumkin 2005). Certainly a number of research questions that would allow us to better understand the sense of place – health relationship have already been raised in this chapter thus far; we now try to elucidate a number of investigative directions, into which many of these research questions may be slotted.

A Way Forward

What we hope to have achieved through our collection of chapters is to make the case for examining not just place but sense of place as an important factor in shaping health outcomes. We derived our original position on this from the conceptually rich population health perspective. But such an all-embracing framework is difficult to move ahead in its entirety. Thus as we have examined some changes to the nature of place and health, we have suggested a more modest agenda. Focusing to be sure on relevant determinants of health in relation to specific health outcomes, we emphasize the physical, psychological and social dimensions of health.

A few examples illustrate a beginning to a plausible research agenda for sense of place-health studies. For physical health, for example, if specific landscapes or local areas have elevated rates of cancer, heart disease, diabetes or some other chronic illness, what is the sense of that place for residents and others? In fact, we know how iconic certain places can become when associated with an adverse event (i.e. Love Canal, Chernobyl, Bhopal) and the impact of place disruption. But what are the configurations of sense of place if, for example, a person resides in Cancer Alley or near Alberta's Oil Sands? Do, for example, the economic rewards (and necessity) outweigh the environmental conditions in the area? For psychological health, we may assert that the sense of place research agenda is potentially very rich. How do the emotions shape sense of place with respect to health? And does a particular configuration of sense of place affect the likelihood of specific emotions intensifying with potential health implications? How does place attachment and disruption affect stress and health? How do perceptions of neighbourhood problems (and opportunities) affect sense of place and health outcomes? Bowling et al. (2006) point to the importance of neighbourhood resources for health. The dimensions of stress – isolation, anxiety, loneliness etc – may also affect sense of place and psychosocial health and should be researched. Indeed these psychological characteristics seem vital in understanding the links between place and health and their complex pathways (see Wen et al. 2006, Warr et al. 2007). For social health, the sense of place agenda is a full one. There exists much research on social capital and cohesion and place and health. How does sense of place affect the different kinds of social capital? Is the presence of one kind or another affected by a particular type of sense of place? And how might this be related to health? What is the significance of perceived neighbourbood disorder on sense of place as well as health? Is sense of disorder itself shaped by sense of place? What is happening in socially excluded neighbourhoods or those where broken windows (Wilson and Kelling 1982, Wallace and Wallace 1998) have not been fixed? And what happens where more inclusion and fixing has occurred? How in these cases does sense of place and health change, if at all?

What does this agenda mean methodologically? As we showed in Chapter 5, qualitative approaches are likely to be vitally important as we sort through the textured and nuanced relationships between sense of place and health. For example, to show how local areas can be stratified yet contribute to identity and well-being in an unequal society, Bolam et al. (2006) treated interviews as text and applied discourse analysis to establish those identities. There seems to be an important role for careful ethnographies as the constituent elements of sense of neighbourhood, neighbourhood characteristics, psychological and behavioural responses and health outcomes are teased out. Given the complexity of place and health, observational and non-verbal and non-textual techniques may be appropriate. Visual methodologies have great potential as do the more traditional techniques of participant observation. In many respects, we may learn a great deal from the suite of methods employed in community studies in both geographical and other settings.

Yet there remains an important role for quantitative approaches to provide broader but generalizable findings on sense of place-health relations. As introduced in Chapter 6, the creation of a survey instrument to measure sense of place and health relations may be the beginning of a much more extensive program of research which can map potential relationships by place, time and population. This would allow an understanding of whether or not differences in this relationship may exist across place size – such as small, medium or large communities, or geographical context – such as rural, sub-urban or urban, for example. Tracking this relationship across time in purposively sampled places would allow an understanding of the inevitable impacts of the changing nature of place/health. How this relationship varies by a wide range of population characteristics, such as: gender, age, health status, or residential longevity, for example, would allow a nuanced understanding of this experience, which could possibly have implications for health promotion planning, for example. The opportunities to add a number of overall sense of place outcome variables to valid and reliable survey instruments measuring social capital, social cohesion, or quality of life, for example, may be another plausible way forward. A mixed-methods approach is one other increasingly common methodological approach used in health studies. Due to the wide range of possibilities available, specific to the operationalization of sense of place and health outcomes, it has great potential for investigating sense of place – health relations.

Recognizing the noted limitations of a social determinants approach to health and a population health framework, we still believe it works as a conceptual foundation for the ongoing examination of sense of place-health relations. As suggested earlier, if health is a means, the numerous social determinants located in the population health model operate as the palette from which we are able to define our narrow empirical research questions and outcomes of interest. Certainly environment is the macro determinant of interest, given that it encompasses both the physical space and the social place characteristics of that space. Some, with singular interest in the social place characteristics may be better guided by the macro determinant of social support. In so doing, we are able to maintain a broad conceptual view of health while simultaneously selecting determinants and their reflective variables to qualitative examine and/or quantitatively measure in a narrow empirical manner, specific dimensions of everyday life and experience. Our ultimate goal, in the end is to contribute to place-based research in order to enhance the understanding environment as a health determinant, which may then be applied to solutions for improved health, such as enhancing the health promoting qualities of places.

References

Adams, P. et al. (2001), 'Place in context: Rethinking humanistic geographies', in P.C. Adams, S. Hoelscher and K.E. Till (eds), *Textures of Place: Exploring Humanistic Geographies* (Minneapolis: University of Minnesota Press).

Amin A. (2002), 'Spatialities of globalisation', *Environment and Planning A* 34(3): 385–399.

Araya R. et al. (2006), 'Perceptions of social capital and the built environment and mental health', *Social Science and Medicine* 62: 3072–3083.

Ash, A. (2002), 'Spatialities of globalization', *Environment and Planning A* 34(3): 385–399.

Asongu, J.J. (2007), 'The future of globalization: An analysis of the impact of political, economical, legal and technological factors on global business', *Journal of Business and Public Policy* 1(4): 1–17.

Bentley, M. (2007), 'Healthy cities, local environmental action and climate change', *Health Promotion International* 22(3): 246–253.

Berkman, L.F. and Glass, T. (2000), 'Social Integration, Social Networks, Social Support, and Health', in Berkman, L.F. and Kawachi, I. (eds), *Social Epidemiology* (New York: Oxford University Press), pp. 137–173.

Bertazzi, P. et al. (1989), 'Ten-year mortality study of the population involved in the Seveso incident in 1976', *American Journal of Epidemiology* 129: 1187–1200.

Billig, M. et al. (2006), 'Anticipatory stress in the population facing forced removal from the Gaza Strip', *Journal of Nervous and Mental Disease* 194: 195–200.

Bolam, B. et al. (2006), 'Place-identity and geographical inequalities in health', *Psychology and Health* 21: 399–420.

Bolland, J. et al. (2005), 'The origins of hopelessness among inner-city African-American adolescents', *American Journal of Community Psychology* 36: 293–305.

Bowling, A. et al. (2006), 'Do perceptions of neighbourhood environment influence health?' *Journal of Epidemiology and Community Health* 60: 476–83.

Brisson, D. and Usher, C. (2005), 'Bonding social capital in low-income neighbourhoods', *Family Relations* 54: 644–53.

Brunner, E. and Marmot, M. (1999), 'Social Organization, Stress and Health', in Marmot, M. and Wilkinson, R.G., *Social Determinants of Health Second Edition* (Oxford: Oxford University Press).

Cantrill, J.G. (1998), 'The environmental self and a sense of place: Communication foundations for regional ecosystem management', *Journal of Applied Communication Research 26*: 301–318.

Capra, F. (2002), *The Hidden Connections* (New York: Doubleday).

Carlson, D. and Chamberlain, E. (2003), 'Social capital, health and health disparities', *Journal of Nursing Scholarship* 35: 325–331.

Chomsky, N. (2000), *Rogue States: The Rule of Force in World Affairs* (Cambridge: South End Publishing).

Clapp, R. et al. (2006), 'Environmental and occupational causes of cancer revisited', *Journal of Public Health Policy* 27: 61–76.

Cohen, D. et al. (2000), 'Broken windows and the risk of gonorrhea', *American Journal of Public Health* 90: 250–256.

—— (2003), 'Neighbourhood physical conditions and premature mortality', *American Journal of Public Health* 93: 467–471.

Couto, E. and Hemminki, K. (2007), 'Estimates of heritable and environmental components of familial breast cancer using family history information', *British Journal of Cancer* 96: 1740–1742.

Danaei, G. et al. (2005), 'Causes of Cancer in the World: Comparative Risk Assessment on Nine Behavioural and Environmental Risk Factors', *Lancet* 366: 1784–1793.

Dauncey, G. (2003), 'A sustainable energy plan for the US', *Earth Island Journal* 18(3): 32–35.

Davis, P.D. (1999), *Ecomuseums: A Sense of Place* (Leicester University Press).

DeSilva, M. et al. (2005), 'Social capital and mental illness', *Journal of Epidemiology and Community Health* 59: 519–527.

Doll, R. (1992), 'Health and the Environment in the 1990s', *American Journal of Public Health* July 82(7): 433–441.

Driscoll, M.E. and Kerchner, C.T. (1999), 'The implications of social capital for schools, communities, and cities: Educational administration as if a sense of place mattered', in J. Murphy and K.S. Louis (eds), *The Handbook of Research on Educational Administration*, 2nd edn. (San Francisco: Jossey-Bass Publishers).

Elliott, S. et al. (2001), 'Mapping health in the Great Lakes areas of concern', *Environmental Health Perspectives* 109 (supp5): 817–826.

Eyles, J. et al. (1993), 'The Social Construction of Risk in a Rural Community: Responses of Local Residents to the 1990 Hagersville (Ontario) Tire Fire', *Risk Analysis* 13(3): 281–290.

Feachem, R.G.A. (2001), 'Globalisation is good for your health, mostly', *British Medical Journal* 323: 504–506.

Ferlander, S. (2007), 'The importance of different forms of social capital for health', *Acta Sociologia* 50: 115–128.

Frankish, J. et al. (1996), *Health impact assessment as a tool for population health promotion and public policy* (Vancouver: Institute of Health Promotion).

Frumkin, H. (2005), 'Health, equity and the built environment', *Environmental Health Perspectives* 113: A290–1.

Fullilove, M. et al. (1998), 'Injury and anomie', *American Journal of Public Health* 88: 924–927.

Gesler, W. (2003), *Healing Places* (Maryland: Rowman and Littlefield).

Gillick, M. (2006), *The Denial of Ageing* (Cambridge: Harvard UP).

Goldberg, P.K. and Pavcnik, N. (2006), 'Distributional Effects of Globalization in Developing Countries'. Revised Draft, October 2006. Accessed from: scid.

stanford.edu/events/IndiaJune2007/DevelopmentResearch/Goldberg%205-17-07.pdf.

Green, S. Harvey, P. and Knox, H. (2005), 'Scales of Place and Networks: An Ethnography of the Imperative to Connect through Information and Communications Technologies', *Current Anthropology* 46(5): 805–826.

Gross, H. and Lane, N. (2007), 'Landscapes of the lifespan', *Journal of Environmental Psychology* 27: 225–41.

Hafkin, N. and Taggart, N. (2001), 'Gender, Information Technology and Developing Countries: An Analytic Study', Academy for Educational Development, for the Office of Women in Development Bureau for Global Programs, Field Support and Research, United States Agency for International Development. Accessed from: http://www.onlinewomeninpolitics.org/beijing12/womenandit.pdf.

Haney, T. (2007), 'Broken windows and self-esteem', *Social Science Research* 36: 968–994.

Hawe, P. and Shiell, A. (2000), 'Social capital and health promotion', *Social Science and Medicine* 51: 871–885.

Hecht, S. (2006), 'A biomarker of exposure to environmental tobacco smoke', *Preventive Medicine* 43: 256–260.

Hemminki, K, Bermejo, J.L. and Försti, A. (2006), 'The balance between heritable and environmental aetiology of human disease', *Nature Reviews Genetics* 7: 958–965.

International Centre for Trade and Sustainable Development (ICTSD) (2006), *Linking Trade, Climate Change and Energy*, ICTSD Trade and Sustainable Energy Series, International Centre for Trade and Sustainable Development, Geneva, Switzerland.

Jutras, S. (2003), 'Allez jouer dehors!', *Canadian Psychology* 44: 257–266.

Kawachi, I. and Wamala, S. (2007), 'Globalization and Health: Challenges and Prospects' in K. Ichiro and S. Wamala (eds), *Globalization and Health* (Oxford: Oxford University Press).

Kotkin, J. and Siegel, F. (2000), *Digital Geography: The Remaking of City and Countryside in the New Economy* (Indianapolis: Hudson Institute).

Kubzansky, L.D. and Kawachi, I. (2000), 'Going to the heart of the matter: Do negative emotions cause coronary heart disease? *J. Psychosom Res* 48: 323–37.

Kunitz, S. (2004), 'Social capital and health', *British Medical Bulletin* 69: 61–73.

Kushner, H.I. and Sterk, C.E. (2005), 'The limits of social capital: Durkheim, Suicide and Social Cohesion', *American Journal of Public Health* 95, 7: 1139–1143.

Labonté, R. and Schrecker, T. (2007), 'Globalization and social determinants of health: The role of the global marketplace' (part 2 of 3) *Globalization and Health* 3: 6.

Labonté, R. and Torgerson, R. (2003), 'Frameworks for Analyzing the Links Between Globalization and Health'. A paper prepared for the Globalization, Trade and Health Group, World Health Organization. Revised December 2003. Accessed from: http://www.globalhealthequity.ca/electronic%20library/Framework%20for%20analysing%20links%20between%20Globalization%20&%20Health%20-%20WHO%201-71%20Dec%2031%202003.pdf.

Lee, K. and Whitman, J. (2003), *Globalization and Health: An Introduction* (Basingstoke: Palgrave).

Lochner, K. et al. (1999), 'Social capital', *Health and Place* 5: 259–570.

Longnecker, M. and Daniels, J. (2001), 'Environmental contaminants as etiologic factors for diabetes', *Environmental Health Perspectives* 109 (suppl): 871–876.

Luke, D. and Harris, J. (2007), 'Network analysis in public health', *Annual Review of Public Health* 28: 69–93.

MacCulloch, R. (2005), 'Income Inequality and the Taste for Revolution', Journal of Law and Economics 48(1): 93–123.

Martin, J.D. and Yidegiligne, H.M. (1998), 'Diabetes mellitus in the First Nations population of British Columbia, Canada', *International Journal of Circumpolar Health* 57 (Supl 1): 335–339.

Massey, D. (ed.) (1994), 'A Global Sense of Place' in *Space, Place and Gender* (Minneapolis: University of Minnesota Press).

Matheson, F. et al. (2006), 'Urban neighbourhoods, chronic stress, gender and depression', *Social Science and Medicine* 63: 2604–2616.

Mathur, K. (1960), 'Environmental factors in coronary heart disease', *Circulation* 21: 684–689.

Menne, B. and Bertollini, R. (2005), 'Health and climate change: a call for action', *British Medical Journal* 331: 1284–1285.

Meyer, J.W., Drori, G.S. and Hwang, H. (2006), 'World Society and the Proliferation of Formal Organization', in G.S. Drori, J.W. Meyer and H. Hwang (eds), *Globalization and Organization: World Society and Organizational Change* (New York: Oxford University Press).

Meyrowitz, J. (2005), 'The Rise of Glocality: New Senses of Place and Identity in the Global Village', in K. Nyiri (ed.), *A Sense of Place: The Global and the Local in Mobile Communication* (Vienna: Passagen Verlag).

Mohan, J. et al. (2004), *Social Capital, Place and Health* (London: Health Development Agency).

Narayan, G. (2007), 'Addressing the Digital Divide: E-Governance and M-Governance in a Hub and Spoke Model', *Electronic Journal of Information Systems in Developing Countries* 31(1): 1–14.

Pampalon, R. et al. (2007), 'Perceptions of place and health', *Social Science and Medicine* 65: 95–111.

Poortinga, W. (2006), 'Social relations or social capital?', *Social Science and Medicine* 63: 255–270.

Public Health Agency of Canada (2007), What determines health? www.phac-aspc.gc.ca/ph-sp/phdd/detrminants/index.html (accessed 10 December 2007).

Raposo, C. et al. (2007), 'Causes of lung cancer', *Medicina Clinica* 128: 390–396.

Relph, T. (2007), 'Spirit of Place and Sense of Place in Virtual Realities', *Techné* 10(3): 17–25.

Roberts, R. and Chen, M. (2006), 'Waste incineration – how big a health risk?' *Journal of Public Health* 28: 261–266.

Rosenlund, M. (2005), *Environmental Factors in Heart Disease* (Stockholm: Karolinska Institute).

Ross, C. (2000), 'Neighbourhood disadvantage and adult depression', *Journal of Health and Social Behavior* 41: 177–187.

Ross, C. et al. (2001), 'Powerlessness and the amplification of threat', *American Sociological Review* 66: 568–591.

Sampson, R. et al. (1997), 'Neighbourhoods and violent crime', *Science* 277: 918–924.

—— (1999), 'Beyond social capital', *American sociological Review* 64: 633–660.

Sampson, R. and Raudenbush, S. (2004), 'Seeing disorder', *Social Psychology Quarterly* 67: 317–342.

Saracci, R. (1997), 'The world health organization needs to reconsider its definition of health', BMJ 314: 1409.

Scholte, J.A. (2005), *Globalization: A Critical Introduction* 2nd edn. (New York: Palgrave Macmillan).

Sergeev, A. and Carpenter, D. (2005), 'Hospitalization rates for coronary heart disease in relation to residence near areas contaminated with persistent organic pollutants and other pollutants', *Environmental Health Perspectives* 113: 756–761.

Smith, R. (2002), 'In search of non-disease', BMJ 324: 883–885.

Sobel, D. (1998), *Mapmaking with Children: Sense of Place Education for the Elementary Years* (Portsmouth: Heinemann.)

Szasz, T. (2007), *Coercion as Cure* (London: Transaction Publishers).

Thapa, S. and Hauff, E. (2005), 'Psychological distress among displaced persons during an armed conflict in Nepal', *Social Psychiatry and Psychiatric Epidemiology* 40: 672–679.

Troung, K. and Ma, S. (2006), 'A systemic review of relations between neighbourhoods and mental health', *Journal of Mental Health Policy and Economics* 9: 137–154.

Wallace, D. and Wallace, R. (1998), *A Plague on your Houses* (London: Verso).

Warr, D. et al. (2007), 'Money, stress and jobs', *Health and Place* 13: 743–756.

Welshman, J. (2006), 'Searching for social capital: Historical perspectives on health, poverty and culture', *The Journal of the Royal Society for the Promotion of Health* 126(6): 268–274.

Wen, M. et al. (2006), 'Objective and perceived neighbourhood environment, individual SES and psychosocial factors and self-rated health', *Social Science and Medicine* 63: 2575–2590.

Wilkinson, D. (2007), 'Local social ties and willingness to intervene', *Justice Quarterly* 24: 185–220.

Wilkinson, R. (1996), *Unhealthy Societies* (London: Routledge).

Williams, D.R. and Stewart, S.I. (1998), 'Sense of place: An elusive concept that is finding a home in ecosystem management', *Journal of Forestry* 96(5): 18–23.

Wilson, J. and Kelling, G. (1982), 'Broken windows', *Atlantic Monthly*: March.

Index